Sergius Dell

Seismic processing and imaging with diffractions

Sergius Dell

Seismic processing and imaging with diffractions

Theory and application

Südwestdeutscher Verlag für Hochschulschriften

Impressum / Imprint

Bibliografische Information der Deutschen Nationalbibliothek: Die Deutsche Nationalbibliothek verzeichnet diese Publikation in der Deutschen Nationalbibliografie; detaillierte bibliografische Daten sind im Internet über http://dnb.d-nb.de abrufbar.
Alle in diesem Buch genannten Marken und Produktnamen unterliegen warenzeichen-, marken- oder patentrechtlichem Schutz bzw. sind Warenzeichen oder eingetragene Warenzeichen der jeweiligen Inhaber. Die Wiedergabe von Marken, Produktnamen, Gebrauchsnamen, Handelsnamen, Warenbezeichnungen u.s.w. in diesem Werk berechtigt auch ohne besondere Kennzeichnung nicht zu der Annahme, dass solche Namen im Sinne der Warenzeichen- und Markenschutzgesetzgebung als frei zu betrachten wären und daher von jedermann benutzt werden dürften.

Bibliographic information published by the Deutsche Nationalbibliothek: The Deutsche Nationalbibliothek lists this publication in the Deutsche Nationalbibliografie; detailed bibliographic data are available in the Internet at http://dnb.d-nb.de.
Any brand names and product names mentioned in this book are subject to trademark, brand or patent protection and are trademarks or registered trademarks of their respective holders. The use of brand names, product names, common names, trade names, product descriptions etc. even without a particular marking in this works is in no way to be construed to mean that such names may be regarded as unrestricted in respect of trademark and brand protection legislation and could thus be used by anyone.

Coverbild / Cover image: www.ingimage.com

Verlag / Publisher:
Südwestdeutscher Verlag für Hochschulschriften
ist ein Imprint der / is a trademark of
AV Akademikerverlag GmbH & Co. KG
Heinrich-Böcking-Str. 6-8, 66121 Saarbrücken, Deutschland / Germany
Email: info@svh-verlag.de

Herstellung: siehe letzte Seite /
Printed at: see last page
ISBN: 978-3-8381-3311-9

Zugl. / Approved by: Hamburg, UHH, Diss., 2012

Contents

Abstract

Reflected and diffracted waves have different nature and role in the applied seismics. Moreover, diffractions itself can be classified in both real seismic diffractions and hypothetical seismic diffractions. The real seismic diffractions are seismic waves which are scattered on small heterogeneities in the subsurface or diffracted at the edges and tips, and recorded as the diffracted part of the whole wavefield. The hypothetical diffractions are mathematical constructions resulting from Huygens principle which helps to correctly image reflected events. These Huygens diffractions build a kernel of seismic reflection imaging, particularly, time migration applies to hyperbolas. The diffraction response in terms of a true Greens' function of the subsurface is good for depth imaging too. The time-migrated data represent pure reflected data with a higher resolution because real seismic diffracted events are collapsed, triplications unfolded and, thus, do not interfere with reflected events.

I present an approach to build the 'partly' time-migrated data. A partly time-migrated gather represents the energy of a hypothetical diffractor collected from several neighboring common-midpoint (CMP) gathers and spread in a considered output gather. The gather has a hyperbolic moveout and is known as the common scatterpoint (CSP) gather. A key concept of the method to generate CSP gathers is a new parametrization of the double square root (DSR) operator with the common-offset apex time. The stacking of the CMP data with the parametrized DSR operator represents a partial time migration by a direct mapping of summed amplitudes into the common-offset operator apex.

The traveltimes of time-migrated reflections can be expressed in terms of the ray propagation in the vicinity of the central image ray. This allows to establish a multiparameter stacking operator to produce time-migrated stacked sections. The multiparameter stacking operator depends on the model-based kinematic wavefield attributes.

Since time-migrated reflections can be obtained by tracing of the image rays down in the subsurface, their kinematic attributes can be used to build smoothed velocity models. Moreover, estimation of kinematic attributes based on the coherence measure is preferable in the time domain due to absence of diffractions and triplications and regularity of the time-migrated gathers.

Although hypothetical diffraction gathers are very preferable to build seismic images, they are not suitable for high-resolution structural imaging. To image objects beyond the classical Rayleigh limit, it is indispensable to use real seismic diffractions. However, the imaging of seismic diffractions is a great challenge in seismic processing because they have weaker amplitudes compared to dominant reflected events. Separation of diffracted from reflected events is frequently used to enhance images of diffractions, and thus to achieve a super-resolution image. I present a method to effectively separate and image diffracted events in the time domain. The approach comprises two steps: attenuation of reflected events by a simultaneous application of the common-reflection-surface-based diffraction operator and a diffraction filter followed by a subsequent velocity analysis in both time and depth domain.

1 Introduction

An explosive point source or a vibration source transmits seismic waves in all directions. During propagation in the subsurface, seismic waves encounter heterogeneities where they may be reflected, refracted, converted or scattered. Reflected or scattered seismic waves propagate back to the surface where they are recorded as seismograms. Seismic processing and imaging is applied to recorded wavefields to create an image of subsurface heterogeneities representing the geological features in the earth. The heterogeneities with the dimensions far above the size of the prevailing wavelength in a seismic signal are referred to as reflectors while the heterogeneities of a size comparable to the prevailing wavelength in the seismic signal are referred to as diffractors. The reflected waves are the kernel of seismic imaging. All conventional processing techniques are tuned to enhance specular reflections. However, reflected waves are not suitable for high-resolution structural imaging, i.e., imaging of features beyond the classical Rayleigh limit of half a seismic wavelength. Typical examples are small-size scattering objects, pinch-outs, fracture corridors, and karst structures. Therefore, a contradiction is given in seismic processing. On the one hand, the diffractions contaminate the data and need to be attenuated to improve reflection imaging. On the other hand, the diffractions are indispensable for high-resolution structural imaging. Therefore, I present two different approaches in this thesis: one technique to attenuate the real seismic diffraction in the prestack data using hypothetical Huygens diffractions and another technique to improve real seismic diffractions. The final aim is an optimal image of the subsurface.

As mentioned above reflection imaging, including the common-reflection-surface (CRS) method, suffers due to the presence of non-specular events in seismic data.

5

Especially when diffractions or triplications are located close to reflections. In such regions the quality of the estimated kinematic wavefield attributes is not sufficient, which affects further seismic applications. For instance, the CRS method provides a smeared stacked section or the velocity model building with Normal Incident Point (NIP) wave tomography may provide an erroneous velocity model. In this context, it is reasonable to extract the kinematic wavefield attributes in the time-migrated domain that contains pure reflections (Tygel et al., 2009). I propose to use CSP gathers which are 'partly' time-migrated data with a hyperbolic moveout based on the distances from the scatterpoint location to a collocated source and receiver. The moveout is dip independent, thus, there is no reflection point smearing in the CSP gathers. The CSP gathers are generated from the CMP gathers by the CSP data mapping (Dell et al., 2012a). The CSP data mapping is based on a new parametrization of the DSR operator for the common offset (CO) operator apex. The parametrized migration operator assigns stacked amplitudes directly into its CO apex while the conventionally migration operator assigns the stacked amplitudes into its ZO apex. The CSP gathers are stacked by an application of a CRS-like stack. The multiparameter stacking technique applied to the time-migrated data not only focuses the time-migrated reflections at zero-offset but also provides a parameter set. The multiparameter stacking operator is based on a hyperbolic approximation of rays propagating in the vicinity of the central image ray which hits the reflector at the Image Incident Point (IIP). Since the multiparameter operator approximates a response of the time-migrated reflector element centered around the IIP, it is called Common-Migrated-Reflector-Element (CMRE) operator.

The CMRE method in the time-migrated domain provides model-based kinematic wavefield attributes while the conventional CRS method determines these attributes on the surface. These model-based parameter can be described in terms of reflector dip and reflector curvature (see e.g. Tygel et al., 2009) as well as the wavefront curvature of the image ray (Dell and Gajewski, 2011). The CMRE stack comprises an automatic CSP stack which represents the high-density velocity analysis and provides an updated time-migration velocity model with improved vertical and lateral resolution. Moreover, the CMRE stack accounts for the neighboring CSPs so that more traces can be involved in the stacking resulting in an enhanced signal-to-noise ratio. Because of an increased reflector resolution and the absence of diffractions, the CMRE stack provides an improved time migration velocity model and hence a highly focused time image.

The propagation of the image ray is determined in terms of the second-order derivative of time-migrated traveltimes with respect to the offset. The image-ray propagation can be described in terms of a hypothetical IIP-wave. The IIP-wave is the wave which starts to propagate in a medium with effective homogeneous properties when a point source explodes at the IIP. Curvatures of IIP-waves can be used to reconstruct a depth velocity model. The velocity model is said to be consistent with the data if all considered IIP-waves focus at zero traveltimes when propagated back into the subsurface. The data vector for the inversion contains curvatures of the IIP-waves which we directly extract from the prestack CSP gathers. The model vector is calculated by dynamic ray-tracing along central image rays. Although the inversion problem is nonlinear and, thus, requires a solution of global nonlinear optimization problem, it is solved iteratively by computing the least-squares solution to the locally linearized problem. The required Fréchet derivatives for the tomographic matrix are calculated with ray perturbation theory.

As mentioned above, reflected waves are not suitable for high-resolution structural imaging, i.e., imaging of features beyond the classical Rayleigh limit of half a seismic wavelength (Khaidukov et al., 2004; Moser and Howard, 2008). Typical examples are small-size scattering objects, pinch-outs, fracture corridors, and karst structures. Imaging and monitoring of these features can be essential for the geological interpretation. The diffracted waves allow us to distinctly detect such local heterogeneities and image them (Landa et al., 1987; Landa and Keydar, 1997). Also diffracted waves indicate the presence of faults and are essential in their investigations (Krey, 1952; Kunz, 1960). Moreover, diffractions can serve as quality control for velocity models in migration methods. Conventional migration methods use only reflections. The velocity model is consistent with the data if seismic events in common-image-gathers (CIG) are flat. However, a velocity model is also consistent with the data if the primary diffractions are focused to points. Velocity analysis based on diffraction focusing may be used to determine migration velocities instead of CIG flatness analysis (Sava et al., 2005). The reflection imaging may also profit from proper imaging of diffractions. Moreover this analysis is performed in the poststack domain with its superior S/N ratio and reduced data volume compared to analysis in the prestack domain.

I introduce an approach for diffraction imaging which combines reflection attenuation and coherent summation of diffracted events (Dell, 2011). The method uses the CRS technique (Tygel et al., 1997). For 2-D media the traveltime t of reflection events is described by three stacking parameters which are commonly referred to as kinematic wavefield attributes: the angle of emergence β_0 of the zero-offset (ZO) ray, the radius of curvature of the normal (N) wave R_N and the radius of curvature of the normal-incidence-point (NIP) wave R_{NIP}. I show that the kinematic wavefield attributes allow to separate reflected and diffracted events in the poststack domain into diffraction-only data. I then perform poststack time-migration velocity analysis on the diffraction-only data. I use the semblance norm as a criterion for optimum migration velocity. The determined velocity model can be used for the update of time-migration velocities as well as for the time-migration of the diffraction-only data. Finally, I image diffraction-only data using the determined velocity model.

The thesis is structured as follows:

A short introduction to the reflection seismic imaging with Huygens's diffraction trajectories as well as the imaging of real seismic diffractions is given in this chapter.

In **Chapter** 2, basics of the Kirchhoff prestack time-migration approach are reviewed and the Common Scatter Point data mapping is introduced. Also an application of the method to synthetic data is shown.

In **Chapter** 3, the basics of hyperbolic traveltime approximation are presented and as an example, the CRS approximation, is briefly reviewed. Also the CMRE stack is introduced. The CMRE represents a new multiparameter stacking operator in the time-migrated domain. It focuses time-migrated reflections at zero-offset. Applications of CRS and CMRE stacking techniques to synthetic data are shown as well.

Chapter 4 reviews the tomographic inversion problem based on the kinematics of reflections, the NIP-wave tomography. The NIP wave tomography is based on the wavefront curvatures of the normal ray. Also I review the image-ray concept and introduce a method to build smoothed depth-migration velocity models based on the wavefront curvature of the image ray, the image ray tomography. The application of the image ray tomography finalizes this chapter.

In **Chapter** 5, I review main properties of diffracted and scattered waves and introduce a technique to image real seismic diffractions in time and depth domain, respectively. The velocity model building uses a semblance based analysis of diffraction traveltimes. Numerical tests show an application of the technique to synthetic data.

In **Chapter** 6, the application of the new imaging approaches to marine field data is presented.

Chapter 7 summaries the results of the thesis.

In **Chapter** 8 I present an outlook of the presented work, namely, extensions of the developed techniques to anisotropic media.

2 Common Scatterpoint Data Mapping

Prestack time migration (PreSTM) still represents the majority of seismic imaging activities in the industry. The reason for this is the speed and robustness of time imaging and its ability to focus seismic events for most geological settings. One of the preferred PreSTM methods is the common offset (CO) Kirchhoff diffraction stack where wavefield imaging is steered by a migration operator based on the double-square-root (DSR) equation. For every CO section, amplitudes are stacked along CO diffraction traveltimes and assigned to the zero offset (ZO) apex time of the migration operator. If the velocity model is correct, the PreSTM leads to flat common-image gathers. Although the PreSTM provides an enhanced reflection resolution in comparison to the raw data, time-migrated seismic data are not suitable for the further applications like stacking velocity analysis or multiparameter stack. Therefore, several methods have been developed to generate 'partly' time-migrated data that are well suitable for further seismic processing.

The frequently used approach in the industry is a CO migration followed by restoration of inverse normal moveout (NMO). The obtained data contain time-migrated reflections with a hyperbolic moveout and are suitable for further applications. Ferber (1994) proposed a migration to multiple offset which presents data with high-fold, bin-centered adjusted, common-midpoint gathers (Ferber, 1994). Bancroft et al. (1998) expanded the method of migration to multiple offset to the equivalent offset method and gave a theoretical explanation for generated gathers (Bancroft et al., 1998). He called these new 'partly' time-migrated gathers common scatterpoint

(CSP) gathers. A CSP gather is similar to a CMP gather, however the moveout is based on the distance from the sources and receiver to the scatter point location. The CSP gather is focused to zero-offset by a NMO-like stacking completes prestack time migration. The above mentioned methods are based on a reformulation of the DSR operator into a single square root. The single square root uses an equivalent offset representing the surface distance from the scatterpoint to a collocated source and receiver.

However, an output CSP gather contains all traces within an aperture which may lead to a significant increase of the data, especially in the 3D case. For instance, starting from an original fold of 10^2 traces for every CMP, the fold in a CSP gather may achieve 10^7 traces using a large midpoint aperture (Ferber, 1994). Moreover, the transformation with the equivalent-offset method uses the RMS velocity determined for the zero-offset stationary point of the time-migration operator and does not take into account the velocity variation, e.g., the offset variation due to non-hyperbolic moveout.

I propose a new technique to generate CSP gathers by a direct mapping of the summed amplitude into the CO apex of the migration operator. The method is based on a new parametrization of the DSR equation with the CO apex time and subsequent stacking with the parametrized operator. The main difference to the above-mentioned methods for CSP data generation is that the output traces in a CSP gather are generated exactly at the locations of the input traces in the CMP gather. That means, the method does not lead to a significant increase of traces in the data making it more suitable for applications in the 3D case. Also the migration velocity is recalculated for every CO sample so that deviations of the velocity determined at ZO are modified to better fit the traveltime.

The main application of the CSP data is time-migration velocity analysis. Because of the increased reflection resolution and absence of diffractions and triplications, the velocity analysis provides an update of migration velocities. However, the CSP gathers can be used for many other complementary applications, e.g., for an attenuation of diffracted multiples or depth-velocity model building with image-ray

11

tomography (see chapter 4).

2.1 Kirchhoff Diffraction Stack

For reflections from continuous interfaces, the physical principle justifying the Kirchhoff diffraction stack states that a reflector can be composed of an sufficiently dense set of point diffractors. The reflective response is the superposition of elementary diffractions from these points and the reflection traveltime surface is the envelope of the elementary diffraction traveltime surfaces. In this case, the elementary diffractions are merely mathematical idealizations, destined to interfere constructively along reflectors and destructively elsewhere due to Huygens's principle. They are abstractions and cannot be observed independently. A real scatter or diffractor is usually a small object in comparison to the dominant wavelength which can be understood as a small reflector with an infinite curvature and an undefined orientation. The characteristic of such a diffractor is that it scatters an elastic wave in all directions. Chapter 5 will be dealing with imaging of real seismic diffractions. In the next two chapters, I assume a point diffractor which is the above-mentioned mathematical idealization.

In a homogeneous model rays are assumed to be straight lines, therefore, the response of a point diffractor can be simply obtained by the application of Pythagoras theorem (Figure 2.1a). I consider a midpoint with coordinates $m = (m_x, m_y)$. Then the source and receiver have coordinates are

$$S = (m_x - h_x, m_y - h_y) \quad \text{and} \quad R = (m_x + h_x, m_y + h_y),$$

where $h_x = h \cos \beta$ is the x-component of the half-offset vector, $h_y = h \sin \beta$ is the y-component of the half-offset vector, and β is the azimuth that is measured with respect to the X axis. The diffraction traveltime t_D associated with the raypath SDR is composed of two branches: the traveltime along the raypath SD and the traveltime along the raypath DR. The distance along the raypath SD from the source to the scatterer in the subsurface is then given by

$$SD = \sqrt{(m_x - h_x)^2 + (m_y - h_y)^2 + z^2}. \tag{2.1}$$

The distance along the raypath DR from the scatterer to the receiver in the subsurface after applying some algebra is given by

$$DR = \sqrt{(m_x + h_x)^2 + (m_y + h_y)^2 + z^2}. \tag{2.2}$$

The traveltime t_D is the sum of Equations 2.1 and 2.2

$$t_D = \frac{1}{v} \left[\sqrt{(m_x - h_x)^2 + (m_y - h_y)^2 + z^2} + \sqrt{(m_x + h_x)^2 + (m_y + h_y)^2 + z^2} \right], \tag{2.3}$$

where v is the constant velocity. Considering the traveltime z/v to be a half zero-offset traveltime $t_0/2$ and inserting it in Equation 2.3 leads to

$$t_D = \sqrt{\frac{t_0^2}{4} + \frac{(m_x - h_x)^2 + (m_y - h_y)^2}{v^2}} + \sqrt{\frac{t_0^2}{4} + \frac{(m_x + h_x)^2 + (m_y + h_y)^2}{v^2}}. \tag{2.4}$$

Equation 2.4 is the well known DSR equation which describes kinematics of the impulse response of a 3D nonzero-offset migration operator applied to 3D prestack data.

In a homogeneous model, the diffraction response of the point diffractor is a hyperboloid of revolution (Figure 2.1b) while in an heterogeneous anisotropic medium, the diffraction response becomes a hyperboloid-like complex and multivalued surface.

The kinematic reflection response of the reflector given by the traveltime surface $t = (\mathbf{m}, \mathbf{h})$ can be reconstructed from the diffraction traveltime surface by computing the envelope surface of all isochrons. For a given source and receiver location, the locus of all possible reflector locations, which satisfies a constant traveltime, is commonly referred to as an isochron surface. For each reflection element in the 3D prestack data there is an isochrone surface which is described by the DSR equation. For convenience, I consider the 3D recording parallel to the inline direction X with zero

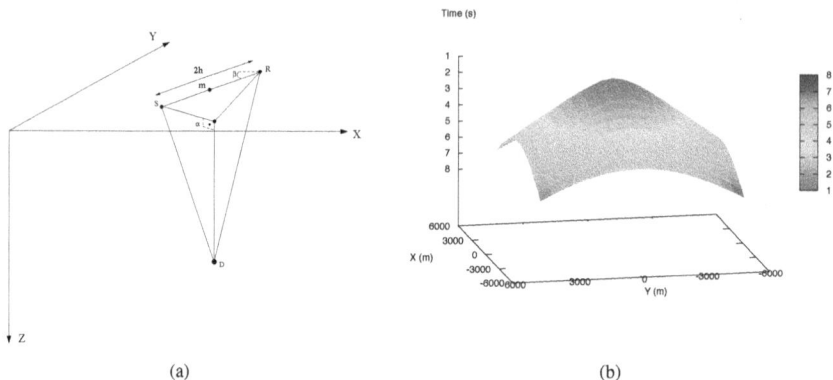

Figure 2.1: Sketch of the 3D geometry (a). Time-migration operator (b). The migration operator is usually parametrized in terms of midpoint coordinates **m**, Euclidean offset $2h$ and offset azimuth β measured with respect to the X axis. Time migration operator depicted for the zero-offset situation. Due to its characteristic shape, the operator is known as Cheops pyramid.

source-receiver azimuth: $\cos \beta = 1, \sin \beta = 0$. Then, after applying some algebra Equation 2.4 can be transformed to

$$\frac{m_1^2}{(vt/2)^2} + \frac{m_2^2}{(vt/2)^2 - h^2} + \frac{z^2}{(vt/2)^2 - h^2} = 1. \tag{2.5}$$

Equation 2.5 describes all isochrons which are a family of ellipsoids. One of the difficulties is to identify correct reflection events which belong corresponding isochrone. An approach to solve this problem is to take one location in the subsurface and determine corresponding traveltimes in the data space. Obviously, these are the same traveltimes as found from the subset of isochrons. In the data space, the surface of these traveltimes is called a diffraction surface. Unfortunately, the diffraction surface tends to cross most reflection events and, if tangent to a reflection event, has a higher curvature. A satisfying solution was proposed by Hagedoorn (1954) who at first stated a relationship between isochrone curves and diffraction curves (Figure 2.2). Instead of reconstructing the reflector at a single output location, Hagedoorn (1954) suggested that the envelope of isochrone surfaces from a reflection

14

event would reconstruct the reflector surface. This statement is the well known Hagedoorn's imaging condition. For a medium with constant velocity the diffraction traveltime surface or Huygens surface t_D of an actual reflection point M_R and the reflection traveltime surface t_R are tangent in the time domain at the stationary point N_R (Figure 2.2). In the same way, the isochrone of a reflection event N_R and the reflector are tangent at M_R in the depth domain. Although the KDS operator is derived using the straight-ray assumption, it can be generalized to a hyperbolic relationship by a Taylor series expansion for an arbitrary medium (Geiger, 2001). The Taylor series expansion assumes the local smoothness of traveltime perturbation in the vicinity of the source and receiver location, but no other assumption about the subsurface. An inhomogeneous isotropic medium is considered so that traveltime variations by perturbing the source and receiver locations are locally smooth. Now we consider that the perturbed source position vector $\mathbf{x}_S = (x_{S_1}, x_{S_2}, x_{S_3})^t$ and receiver position vector $\mathbf{x}_R = (x_{R_1}, x_{R_2}, x_{R_3})^t$ can be expressed in term of variations $\Delta \mathbf{x}_S$ and $\Delta \mathbf{x}_R$ in the unperturbed source and receiver positions as $\mathbf{x}_S = \mathbf{x}_{S_0} + \Delta \mathbf{x}_S$ and $\mathbf{x}_R = \mathbf{x}_{R_0} + \Delta \mathbf{x}_R$. The total traveltime from the source to the point-diffractor location to the receiver is comprises two components. The square of the traveltime from the source to a fixed diffractor location can be expressed as the hyperbolic traveltime expansion about perturbations in the source and receiver location (see Vanelle and Gajewski, 2002).

$$t^2(\mathbf{x}_D, \mathbf{x}_S) = \left(t_{0_{SD}} - \mathbf{p}_{S_0} \Delta \mathbf{x}_S^t\right)^2 - t_{0_{SD}} \Delta \mathbf{x}_S \mathbf{S}_0 \Delta \mathbf{x}_S^t + O(3) \tag{2.6}$$

where $t_{0_{SD}}$ is the traveltime from S_0 to the diffractor, $\mathbf{p}_{S_0}$ is the slowness vector at the unperturbed source position $\mathbf{x}_{S_0}$ defined as

$$p_{S_0 i} = -\left. \frac{\partial t}{\partial x_{S_i}} \right|_{R_0, S_0} \tag{2.7}$$

and $\mathbf{S}_0$ is second-order derivative matrix at the unperturbed source position $\mathbf{x}_{S_0}$ defined as

$$S_{ij} = N_S^R = \left. \left(\frac{\partial^2 t}{\partial (x_S)_i \partial (x_S)_j} \right)_{i,j=1,2} \right|_{R_0, S_0} \tag{2.8}$$

The square of the traveltime from the fixed diffractor location to the receiver also can

15

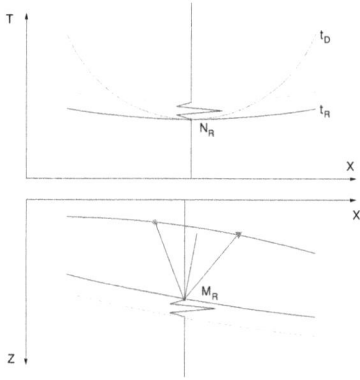

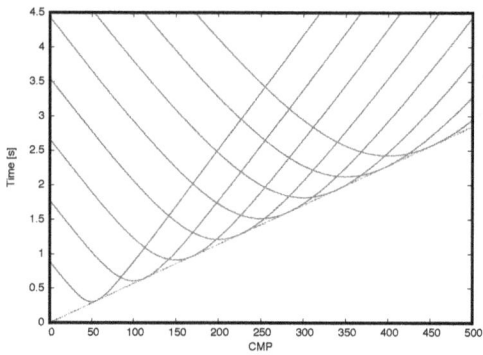

Figure 2.2: Hagedoorn's imaging condition. (a) The Huygens curve $t_D(M_R)$ of a point on the reflector M_R is tangent to the reflection traveltime curve $t_R(N_R)$ at point N_R in the time domain. (b) The reflection traveltime curve (green line) perfectly coincides with the envelope (blue line) of diffraction curves (red lines). The reflector is inclined.

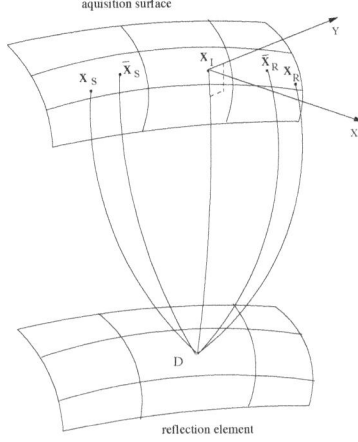

Figure 2.3: A generalized isotropic medium with only one additional assumption that traveltime variations by perturbing the source and receiver locations are locally smooth. The image ray starts normally to the acquisition surface and hits the reflector element at the IIP, denoted by P, which coincides with the location of the point diffractor.

be expressed as the hyperbolic traveltime

$$t^2(\mathbf{x}_R, \mathbf{x}_D) = \left(t_{0_{DR}} + \mathbf{p}_{R_0} \Delta \mathbf{x}_R^t\right)^2 + t_{0_{DR}} \Delta \mathbf{x}_R \mathbf{R}_0 \Delta \mathbf{x}_R^t + O(3) \tag{2.9}$$

with the traveltime $t_{0_{DR}}$ from the diffractor to R_0, the slowness vector where $\mathbf{p}_{R_0}$ at the unperturbed receiver position $\mathbf{x}_{R_0}$ defined as

$$p_{R_i} = \left. \frac{\partial t}{\partial x_{R_i}} \right|_{R_0, S_0} \tag{2.10}$$

and the second derivative matrix $\mathbf{R}_0$ at the unperturbed receiver position $\mathbf{x}_{R_0}$ defined as

$$R_{ij} = N_R^S = \left. \left(\frac{\partial^2 t}{\partial (x_R)_i \partial (x_R)_j} \right)_{i,j=1,2} \right|_{R_0, S_0} \tag{2.11}$$

The sum of these component traveltimes yields the total diffraction traveltime

$$t(\mathbf{x}_R, \mathbf{x}_S) = t(\mathbf{x}_D, \mathbf{x}_S) + t(\mathbf{x}_R, \mathbf{x}_D) \tag{2.12}$$

17

which can be expressed as a generalized DSR equation (Geiger, 2001).

$$t(\mathbf{x}_R, \mathbf{x}_S) = \sqrt{\left(t_{0_{SD}} - \mathbf{p}_{S_0}\Delta\mathbf{x}_S^t\right)^2 - t_{0_{SD}}\Delta\mathbf{x}_S \mathbf{S}_0 \Delta\mathbf{x}_S^t} +$$
$$+ \sqrt{\left(t_{0_{DR}} + \mathbf{p}_{R_0}\Delta\mathbf{x}_R^t\right)^2 + t_{0_{DR}}\Delta\mathbf{x}_R \mathbf{R}_0 \Delta\mathbf{x}_R^t} + O(3)$$

(2.13)

The slownesses and the second-order derivatives are the inverses of the apparent velocity and the curvature of the wavefront at the unperturbed source and receiver locations.

The point diffractor lies in the plane $x_3 = 0$ (see Figure 2.3). The distances to the source and receiver are measured from the origin, and the midpoint vector and offset vector are half the sum and half of the difference of these distances, respectively. For an arbitrary zero-offset ray emerging at the origin $\mathbf{x}_{S_0} = \mathbf{x}_{R_0} = (0,0,0)$, the zero-offset apparent slowness vectors are equal $\mathbf{p}_{R_0} = -\mathbf{p}_{S_0} = \mathbf{p}_0$ and the zero-offset apparent curvature $\mathbf{R}_0 = -\mathbf{S}_0 = \mathbf{P}_0$ with $P_{0_{11}} = P_{0_{22}} = P$ and $P_{0_{12}} = P_{0_{21}} = 0$, $t_D = t(\mathbf{x}_R, \mathbf{x}_S)$, and $t/2 = t_{0_{DR}} = t_{0_{SD}}$. Inserting these into Equation 2.13 yields

$$t_D = \sqrt{\left(\frac{t_0}{2} + p\left[m_x + m_y - h_x - h_y\right]\right)^2 + \frac{t_0}{2}P\left[(m_x - h_x)^2 + (m_y - h_y)^2\right]} +$$
$$+ \sqrt{\left(\frac{t_0}{2} + p\left[m_x + m_y + h_x - h_y\right]\right)^2 + \frac{t_0}{2}P\left[(m_x + h_x)^2 + (m_y + h_y)^2\right]}$$

(2.14)

When one considers the expansion point to be the emergence location of the image ray, then the horizontal projection of the slowness vector is zero, i.e., $p = 0$. The image ray is a propagation path of the energy from the point diffractor location to the apex of the migration operator Hubral (1977). In this case Equation 2.14 simplifies to

$$t_D = \sqrt{\frac{t_0^2}{4} + \frac{(m_x - h_x)^2 + (m_y - h_y)^2}{v_{mig}^2}} + \sqrt{\frac{t_0^2}{4} + \frac{(m_x + h_x)^2 + (m_y + h_y)^2}{v_{mig}^2}}$$

(2.15)

with

$$v_{mig} = \sqrt{\frac{2}{t_0 P}}.$$
(2.16)

The parameter v_{mig} is a fitting parameter commonly referred to as time-migration velocity. v_{mig} depends on the wavefront-curvature of the image ray .

Equation 2.15 is a generalized diffraction traveltime expression for an arbitrary inhomogeneous isotropic medium (Geiger, 2001). The hyperbolic traveltime expression is expanded in the vicinity of the zero-offset image ray, i.e., traveltime t_0 is measured along this ray. The traveltime is also described by a single parameter that corresponds to the wavefront curvature at the image ray emerged at the location $m_0, h = 0$ and with traveltime $t_0/2$. The extraction of this parameter is usually complicated since the exact solution requires the constant velocity medium near the image ray location. However, a good initial solution is possible in a layer cake medium near the image ray location. For such media, root-mean-square (RMS) velocities can be used instead of time-migration velocities (Dix, 1955). For 1D media the RMS velocity is given by

$$v_{rms}^2 = \frac{1}{t_0} \int_0^{t_0} v^2(t)\,dt$$
(2.17)

and it equals the normal moveout (NMO) velocity. It is a common approach to use NMO velocities instead of RMS velocities for time migration. The velocity $v(t)$ represents the subsurface velocity along the normal ray. Thus, the NMO velocity in 1D media constitutes an integral velocity of the overburden of the reflection point. In general, the quantity of the second-order expansion, v_{NMO}, cannot be determined exactly from the data by a single coherence analysis as this yields a best fit quantity v_{stack}, stacking velocity. The stacking velocity is attached to the stationary point of the migration operator in contrast to the migration velocity which is defined at the apex of the time-migration operator. In case of horizontally layered medium, the stacking velocities coincides with the time-migration velocities.

However, the real subsurface exhibits very complex velocity distributions and it is often desired to migrate far offsets. In this case, RMS velocities should be replaced with time-migration velocities determined at the apex time of the time-migration operator. The time-migration velocities control the time-migration operator providing its best-fit to the data. Within the hyperbolic assumption we can consider the CMP stacking velocities to be a very good first approximation of time-migration velocities. For complex media, time-migration velocities usually deviate from the stacking velocities leading to an unfocused time image. Therefore in the industry a velocity update is frequently applied to correct the initial migration velocities. The conventional time-migration velocity update can be carried out as an iterative approach based on residual moveout analysis, a scanning routine, or inverse NMO route (see, e.g., in Robein, 2010). The RMO routine and the inverse NMO route usually start with an initial velocity field using the stacking velocities determined by means of a conventional stacking velocity analysis. Then, selected image gathers are constructed using prestack time migration which usually show residual moveout. Afterward, the residual moveout can be parametrized by a function of one or two parameters in order to fit the residual non-flatness of the image gathers. The search procedure is based on a coherence measure and is usually strongly controlled by the user. Finally, the image gathers are stacked for the time-migrated section. In the inverse NMO route, an inverse NMO correction is applied to selected image gathers using the initial velocities. Here, the assumption is made that the inverse NMO correction returns the image gathers back to the situation before the prestack time migration. However, this assumption is valid only for horizontally layered media. Both techniques are usually carried out on a rather coarse grid, i.e., the time-migration velocities have to be interpolated on the migration grid. The scanning approach tests for a set of velocities or velocity functions at the same time. The best result is determined on the basis of image-gather flatness and interpretative criteria. The approach is rather compute-intensive as several full prestack time migrations have to be performed. In the next section I will present another approach to estimate time-migration velocities using CSP gathers.

Let me sum up main ideas of Kirchhoff diffraction stacking. Each point of a sufficiently dense grid in the target area is considered to be a potential point diffractor. The diffraction traveltime surface is calculated independently for any of these points with the time-migration operator given by the DSR equation (Equation 2.4). The amplitudes are stacked along the diffraction traveltime surface and the migration output is assigned to the ZO apex of the migration operator. In the next section I will describe a time-migration like method where stacked amplitudes assigned to the CO apex of the migration operator.

2.2 Common Scatterpoint Data Mapping

Common Scatterpoint Data Mapping is a time-migration like method without the collapsing scattered energy in the offset direction. For the sake of simplicity, I firstly consider the 2-D case. The 3-D case will be generalized on the end of this section. The traveltime surface of a single scatter point at (m_0, t_0) is given by the 2-D DSR equation

$$t_D = \sqrt{\frac{t_0^2}{4} + \frac{(m-h)^2}{v_{mig}^2}} + \sqrt{\frac{t_0^2}{4} + \frac{(m+h)^2}{v_{mig}^2}} \tag{2.18}$$

and is known as the Cheops pyramid. The goal of the CSP data mapping is to generate 'partly' time-migrated gathers that are suitable for further seismic processing. The CSP gather building process can be defined as a collection of energy along a diffraction traveltime surface in 3-D space of midpoint, offset and time (m, h, t) (Figure 2.4a) and the scattering of this energy in a 2-D space of offset and time (h, t) along a hyperbolic path (Figure 2.4b). The hyperbolic path also represents the CO apexes of the time-migration operator and reads as

$$t_{apex} = \sqrt{t_0^2 + \frac{4h^2}{v^2}}, \tag{2.19}$$

where t_0 is the image time that corresponds to the ZO operator apex, h is half source-receiver offset and v is the migration velocity. A CMP gather that is located at the scatter point (m_0) intersects the Cheops pyramid on a hyperbolic path, which coincides with the apexes of the traveltime surface (Figure 2.4). For this CMP all the scattered energy will be focused along the hyperbolic path (Figure 2.4b). The intersections of all other CMP gathers have non-hyperbolic paths. For these CMPs all the scattered energy will be mispositioned.

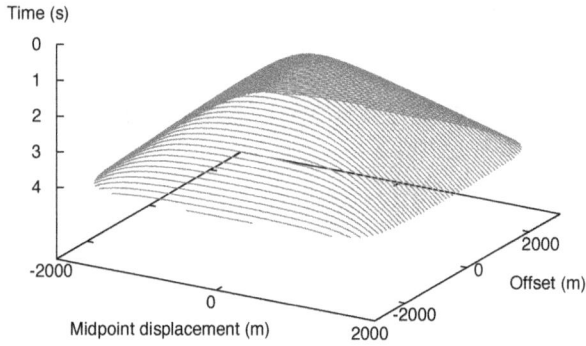

(a)

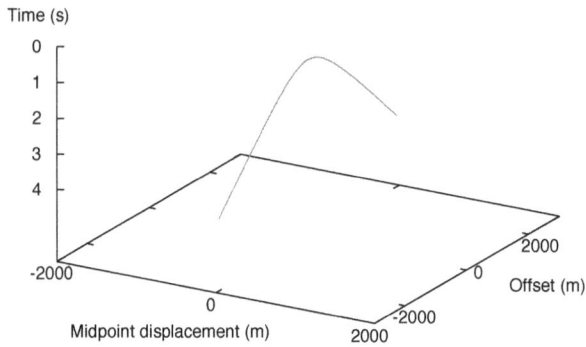

(b)

Figure 2.4: (a) The none-zero traveltime surface $t(x_m, h)$ for a single scatter point. The traveltime surface is known as the Cheops pyramid. (b) The CSP gather. The CSP gather formed by collapsing the Cheops pyramid to a hyperbola in the $x_m = 0$ plane. The apex of the hyperbola is at $(x_m = 0, h = 0)$. The line coincides with summations trajectories for the CSP operator.

To assign the migration output to the CO apex of the migration operator we propose parametrize the DSR equation with its CO apex time as followed:

$$t = \sqrt{\frac{t_0^2}{4} + \frac{m^2 - 2mh + h^2}{v^2}} + \sqrt{\frac{t_0^2}{4} + \frac{m^2 + 2mh + h^2}{v^2}}$$

$$t = \sqrt{\frac{t_0^2}{4} + \frac{h^2}{v^2} + \frac{m^2 - 2mh}{v^2}} + \sqrt{\frac{t_0^2}{4} + \frac{h^2}{v^2} + \frac{m^2 + 2mh}{v^2}}$$

$$t = \sqrt{\frac{1}{4}\left(t_0^2 + \frac{4h^2}{v^2}\right) + \frac{m(m - 2h)}{v^2}} + \sqrt{\frac{1}{4}\left(t_0^2 + \frac{4h^2}{v^2}\right) + \frac{m(m + 2h)}{v^2}}.$$

With equation 2.19 one obtains:

$$t = \sqrt{\frac{t_{apex}^2}{4} + \frac{m(m - 2h)}{v^2}} + \sqrt{\frac{t_{apex}^2}{4} + \frac{m(m + 2h)}{v^2}}. \tag{2.20}$$

Please note, that the velocity in the latter equation is parametrized for the CO apex time t_{apex} but belongs to ZO apex time of the migration operator t_0. Thus, the parametrized DSR equation is finally given by

$$t = \sqrt{\frac{t_{apex}^2}{4} + \frac{m(m - 2h)}{v(t_0)^2}} + \sqrt{\frac{t_{apex}^2}{4} + \frac{m(m + 2h)}{v(t_0)^2}}. \tag{2.21}$$

To find the velocity $v(t_0)$, which corresponds to t_{apex} a search procedure is performed (Figure 2.5). A similar procedure is also applied for the generation of CRS supergather (Baykulov and Gajewski, 2009). All t_0 traveltimes within the time window $[0;t]$ and a range of velocities $[v_{min}, v_{max}]$ are tested for an event $A(t_A, h_A)$ to determine the best fitting hyperbola. The hyperbolas are computed using Equation 2.19. After the time t_0' which corresponds to the minimum deviation between the computed and the actual time for sample $A(t_A, h_A)$ is found, velocity which belongs to t_0 is determined.

Figure 2.6 compares the principle of the time migration and the data mapping.

The CSP data mapping forms the CSP gathers directly from the input CMP gathers

23

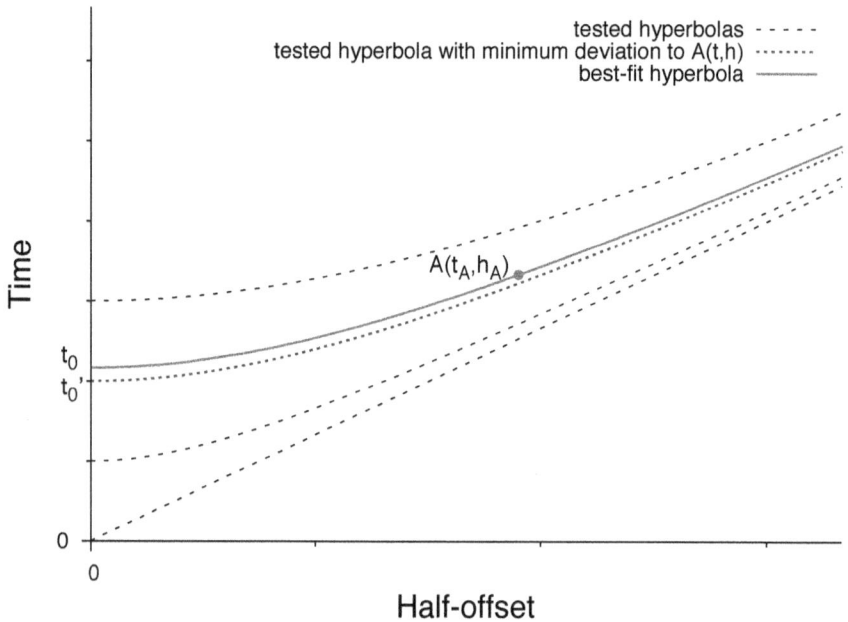

Figure 2.5: Testing traveltime curves to find the best-fit hyperbola for the sample $A(t_A, h_A)$. The sample $A(t_A, h_A)$ corresponds to an apex of the CO migration operator. All t_0 traveltimes within the time window $[0; t]$ and a range of velocities are tested. The time t_0' corresponds to the minimum deviation between the computed and the actual time for sample $A(t_A, h_A)$. This time is assumed to be the searched for ZO apex time. The velocity corresponding to this time is then assumed to be the searched for $v(t_0)$ velocity. The tick interval of the time axis corresponds to the time sample rate of the data. This figure is kindly provided by Dr. Mikhail Baykulov.

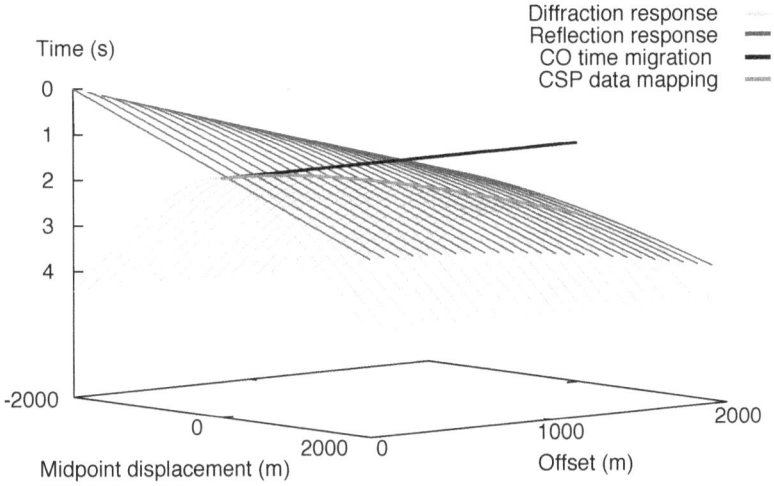

Figure 2.6: The figure compares the principle of time migration and data mapping for a homogeneous model with a dipping reflector. The reflection response is depicted in blue. The migration operator for CMP 250 is depicted in cyan. The migration output is assigned to the ZO operator apex for every CO section (black line). The mapped output is assigned to the CO operator apex for every CO section (red line).

with initial time-migration velocities. As initial time-migration velocities I use the stacking velocities that are obtained as a byproduct after an automatic CMP stack of the CMP gathers. The whole process is performed automatically and does not require any manual picking of the velocities. CSP data represent 'partly' time-migrated data since the scattered energy is not focused to zero offset. An application of NMO-like stacks to the CSP data leads to the time-migrated image.

Now I am going to show, that the stacking parameters extracted from single CSP gathers are related to migration velocities. Since the CSP gathers represent time-migrated data, we can expand the time-migrated traveltimes in the vicinity of the image ray

$$T(h)^2 = (T_0)^2 + \frac{T_0}{2}\frac{\partial^2 T}{\partial h^2}h^2,$$ (2.22)

where h is the half-offset. The equation above I rewrite as

$$T(h)^2 = (T_0)^2 + \frac{h^2}{v^2},$$ (2.23)

where

$$v^2 = \frac{2}{T_0 \frac{\partial^2 T}{\partial h^2}}.$$ (2.24)

Taking into account that the derivative $\partial^2 T/\partial h^2$ describes the wavefront curvature of the image ray and the expression for time-migration velocity given by Equation 2.16 provides

$$v = \sqrt{\frac{2}{T_0 \frac{\partial^2 T}{\partial h^2}}} \equiv \sqrt{\frac{2}{t_0 P}} = v_{mig}.$$ (2.25)

The time-migration velocities can be estimated from CSP gathers by conventional stacking velocity analysis.

Now I generalize the CSP data mapping to the 3-D case. The 3-D time-migration operator is given by Equation 2.15. This operator defines a 5-D subsurface. However, we aim to find the minimum of this subsurface with respect to the offset vector $\mathbf{h}$. For this purpose, we calculate the first derivatives of the traveltime subsurface with respect to offset vector $\mathbf{h} = (h_x, h_y)$ and set it equal to zero. After applying some algebra, we obtain for the 3-D case the expression for the minimum of the traveltime subsurface

$$t_{apex}(h_x, h_y) = \sqrt{t_0^2 + \frac{4(h_x^2 + h_y^2)}{v^2}}. \tag{2.26}$$

Inserting this equation into Equation for the 3-D migration operator given by Equation 2.15 provides eventually the 3-D CSP mapping operator. The operator reads as

$$t = \sqrt{\frac{t_{apex}^2}{4} + \frac{m_x(m_x - 2h_x) + m_y(m_y - 2h_y)}{v(t_0)^2}} + \sqrt{\frac{t_{apex}^2}{4} + \frac{m_x(m_x + 2h_x) + m_y(m_y + 2h_y)}{v(t_0)^2}}, \tag{2.27}$$

where t_{apex} does not define the hyperbolic path any more but a hyperbolic surface.

2.3 Synthetic example

In order to verify the proposed method, it was applied to a simple synthetic model. Figure 2.7a displays a stacked section of a synthetic model containing five layers and four small lenses which simulate diffractors. The velocity within the layers is constant. The velocity in the first layer is 1500 m/s, in the second layer 1580 m/s, in the third layer 1690 m/s, in the fourth layer 1825 m/s and in the fifth layer 2000m/s. Four small lenses with a lateral extension of 200 meters in the fourth layer produce diffractions. I used Seismic Un*x to generate synthetic seismograms with the Gaussian beam method. A Ricker-wavelet with a prevailing frequency of 25 Hz was used. Random noise with a S/N of 20 was added to the synthetic data. Figure 2.7 shows the model and the corresponding CRS stacked section.

Figure 2.8 shows CMP and CSP gathers. Diffractions are indicated with red arrows. The CMP gather (a) and CSP gather (b) are located at 1500 m directly above the diffraction apex. Here, the diffraction is present because of the constructively interference (see Figure 2.4). The CMP gather (c) and CSP gather (d) are located at 1800 m, offset to the diffraction apex. In the CSP gather located offset with respect to the apex time diffractions are absent because of destructive interference. The moveout in the CSP gathers seems to coincide with the moveout in the CMP gathers.

27

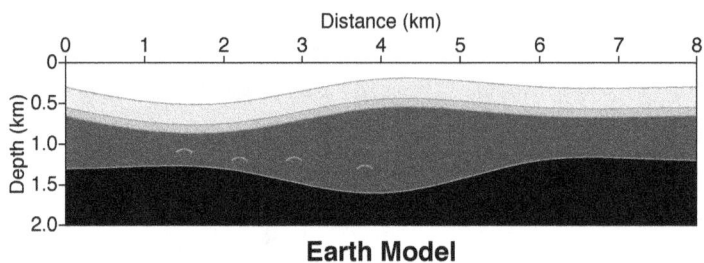

Earth Model

(a)

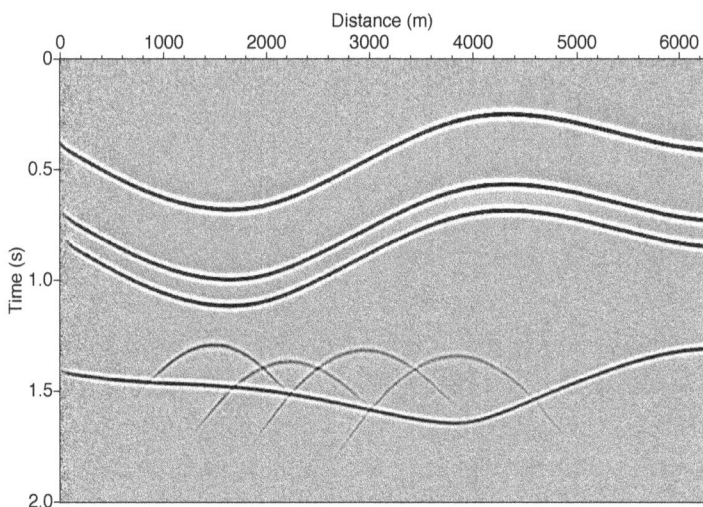

(b)

Figure 2.7: (a) The geological model contains five layers and four small lenses
which simulate diffractors. (b) CRS stack of CMP gathers. The CRS
stacked section shows both reflected and diffracted events. The latter
ones generate conflicting dip situations with the last reflection.

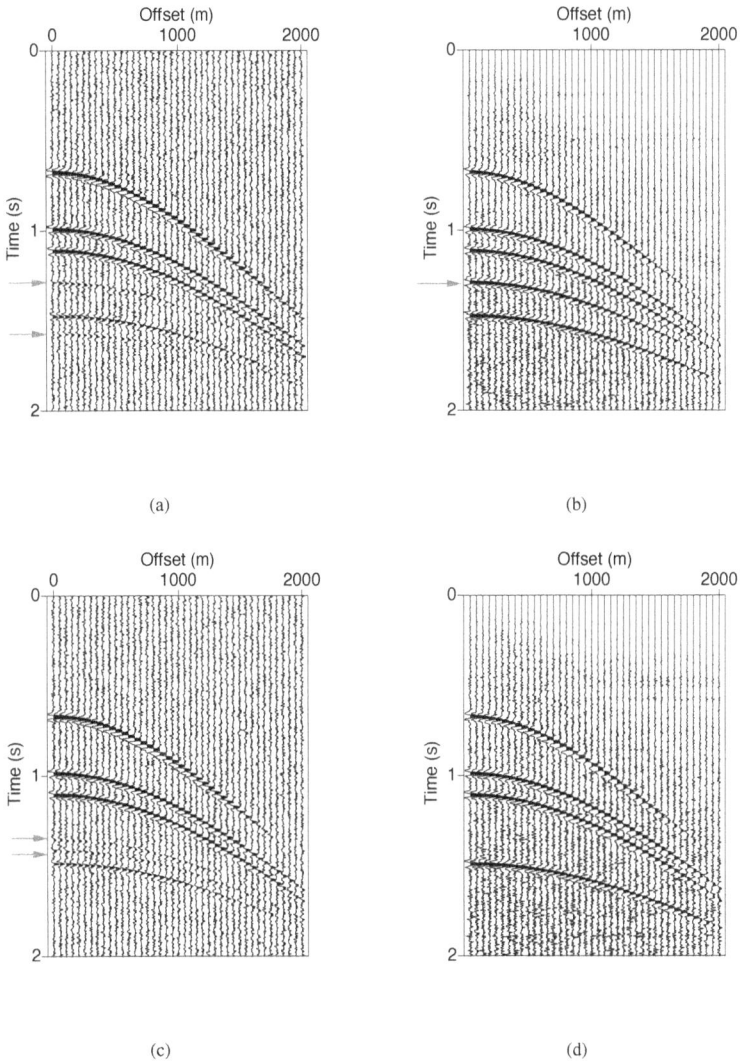

Figure 2.8: CMP (left side) and CSP (right side) gathers. Diffractions are indicated with red arrows. The CMP gather (a) and CSP gather (b) are located at 1500 m directly above the diffraction apex. Here, the diffraction is present because of the constructively interference. The CMP gather (c) and CSP gather (d) are located at 1800 m, offset to the diffraction apex. In the CSP gather located offset with respect to the apex time diffractions are absent because of destructive interference.

However, the moveout in the CSP gathers is based on the distances from a collocated source and receiver to a scatter point location while the moveout in the CMP gather is based on the offsets which are distances from a collocated source and receiver to a reflection point. The similarity is caused by a relative flat character of reflecting interfaces where the reflection point is similar for all offset and also coincide with the scatter point location. For more curved or dipping reflectors the moveout in the CSP gathers can considerably deviate from the moveout in the CMP gathers.

Figure 2.9 shows velocity spectra extracted from CMP and CSP gathers. In the CMP-based velocity spectra, the diffractions provide well focused events. These events are marked by white arrows. The contribution of diffracted events affects the velocity determination for reflections. On the contrary, the velocity spectra estimated from the CSP gathers are not contaminated through such type of events resulting in a better determination of time-migration velocities.

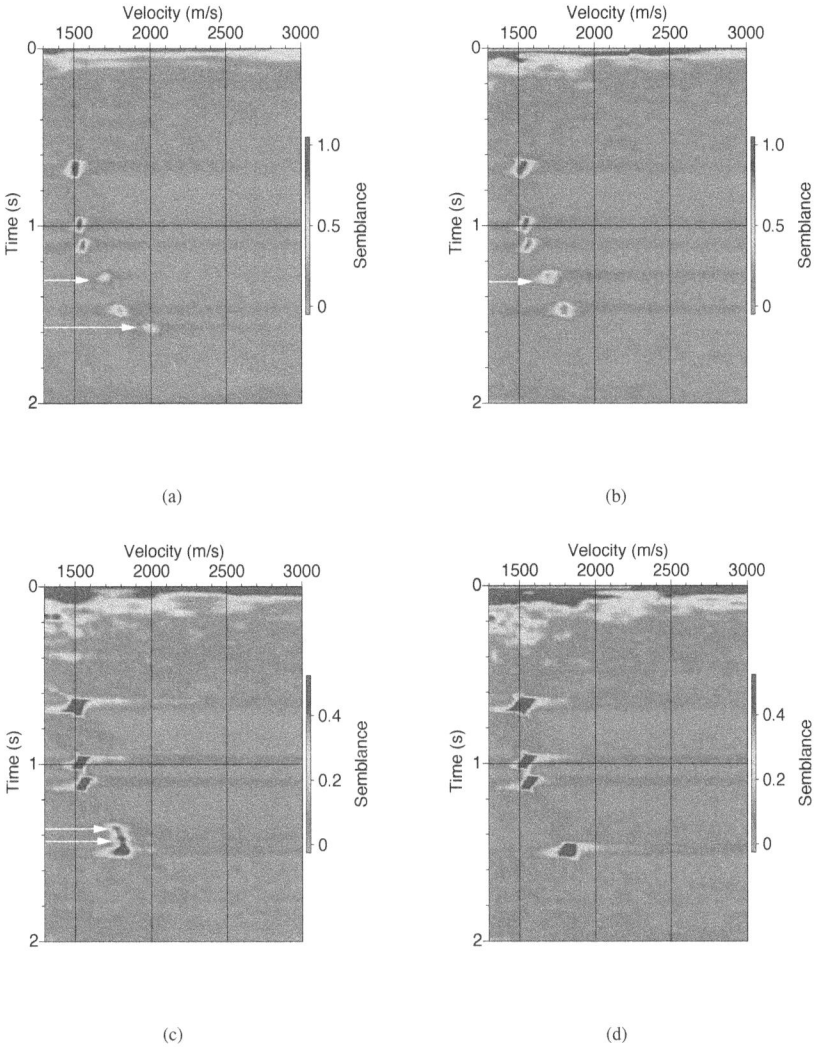

Figure 2.9: The corresponding velocity spectra for CMP (left side) and CSP (right side) gathers displayed in Figure 2.8. In CMP-based velocity spectra, diffractions provide well focused events. These events are marked by white arrows. The contribution of diffracted events affects the velocity determination for reflections. The velocity spectra estimated from the CSP gathers are not contaminated through such type of events resulting in a better determination of time-migration velocities.

31

2.4 Concluding remarks

I have developed a new method that allows to map the CMP gathers into CSP gathers. A 2-D CSP gather collects all scattered energy from a 3-D data volume (m,h,t) within an aperture and redistributes this energy onto a 2-D data volume (h,t) along a hyperbolic path. If a scatter point is exactly at the considered input CMP location, its scattered energy is constructively stacked along this hyperbolic path. Energy from scatter points offset to this CMP location is canceled by destructive interference. The method is based on the principles of Kirchhoff prestack time migration and uses a new parametrization of the DSR equation with CO apex time. The main advantage of the presented method in comparison to other present method is that the summed amplitude is directly mapped into the CO apex of the mapping operator. The time domain formulation of the data mapping allows CSP gathers to be formed at arbitrary locations, i.e, data can be regularized. Also the CSP gathers are very suitable for many complementary applications, e.g., time-migration velocity model building, image-ray tomography, and multiple attenuation.

In the next chapter, I discuss how to approximate the 'partly' time-migrated traveltimes. At the begin of the section, I review basics of the hyperbolic traveltime approximation with emphasize on the CRS approximation. Then I introduce the CMRE approximation which represents a CRS-like approximation of the time-migrated reflections and show an application to a synthetic data example.

3 Traveltimes in the unmigrated and time-migrated domain

In the previous chapter I have presented a method to produce time-migrated data, CSP gathers. Since the CSP data represent 'partly' time-migrated data, they should be stacked to zero-offset. For this purpose, it is necessary to establish a stacking operator. In general, the stacking operator may depend on a single stacking parameter as in the conventional NMO stack (Mayne, 1962) or several stacking parameter as in the CRS stack (Jäger et al., 2001). The CRS operator depends on three stacking parameters and is based on the traveltime expansion in the vicinity of the normal ray. I suggest to use a CRS-like stacking operator to focus time-migrated reflections to zero-offset. The new operator also depends on three stacking parameters and is based on the traveltime expansion in the vicinity of the image ray. Since the new operator approximates the response of a whole migrated-reflector element, I call it Common-Migrated-Reflector-Element operator.

3.1 Hyperbolic traveltime approximation

The CRS method applied to the unmigrated data approximates reflection traveltime in the vicinity of the central normal ray. The traveltime approximation in the vicinity of the central ray is based on paraxial ray theory (Červený, 2001). It describes the 3D traveltime perturbation of nearly rays. This perturbation is expressed in terms of three-component dislocation vectors for the start point and end point of

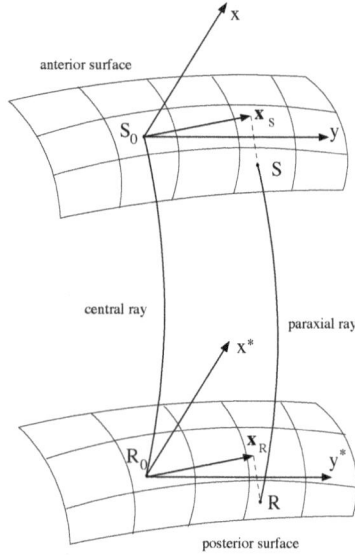

Figure 3.1: Definition of the endpoint coordinate system after Bortfeld (1989). The
vector $\mathbf{x}_S(\mathbf{x}_R)$ is the projection of the paraxial point $S(R)$ on the tangent
plane to the anterior (posterior) surface at $S_0(R_0)$

the nearly rays. The traveltime of these rays is given by a parabolic expression.
However it can be formulated as a hyperbolic expression (Ursin, 1982). Another
parabolic formulation of the traveltime for an arbitrary layered medium with costant
velocity was established by Bortfeld (1989) and extended by Hubral et al. (1992)
to variable velocity within the layers. The perturbed ray endpoints are in both
cases confined to curved surfaces and are described by two-components vectors.
The hyperbolic traveltime approximation based on Bortfeld's formalism can also be
derived and expanded for an arbitrary ray (Schleicher et al., 1993). In the following
an arbitrary fixed ray connecting a source at point S and a receiver at point R in a 3D
inhomogeneous medium is considered (Figure 3.1).

This ray is referred to as central ray. The initial point S and end point R of a
paraxial ray in the close vicinity of the central ray $S_0(R_0)$ lie on the anterior and
posterior surfaces. Following the formalism of Bortfeld tailored to three-dimensional

propagation, the 4×4 propagator matrix $\underline{\mathbf{T}}$, given as

$$\underline{\mathbf{T}}(R,S) = \begin{pmatrix} \mathbf{A} & \mathbf{B} \\ \mathbf{C} & \mathbf{D} \end{pmatrix}$$

describes a first-order relationship

$$\begin{cases} \mathbf{x}_R = \mathbf{A}\,\mathbf{x}_S + \mathbf{B}\,\mathbf{p}_S, \\ \mathbf{p}_R - \mathbf{p}_{R_0} = \mathbf{C}\,\mathbf{x}_S + \mathbf{D}\left(\mathbf{p}_S - \mathbf{p}_{S_0}\right), \end{cases} \quad (3.1)$$

where $\mathbf{A}, \mathbf{B}, \mathbf{C}, \mathbf{D}$ denote constant 2×2 sub-matrices which characterize the central ray

$$A_{ij} = \left.\frac{\partial x_R}{\partial x_S}\right|_{R_0,S_0}, \quad B_{ij} = \left.\frac{\partial x_R}{\partial p_S}\right|_{R_0,S_0}, \quad (3.2)$$

$$C_{ij} = \left.\frac{\partial p_R}{\partial x_S}\right|_{R_0,S_0}, \quad D_{ij} = \left.\frac{\partial p_R}{\partial p_S}\right|_{R_0,S_0}. \quad (3.3)$$

$\mathbf{x}_S$ and $\mathbf{x}_R$ are the dislocation vectors of the paraxial ray with respect to the central ray and $\mathbf{p}_S$ and $\mathbf{p}_R$ denote the corresponding deviations of the slowness vectors. The parabolic traveltime for a ray SR paraxial to a given central ray $S_0(R_0)$ can be expressed as a function of its coordinates at the anterior ($\mathbf{x}_S$) and posterior surface ($\mathbf{x}_R$), respectively, in a second-order approximation by

$$t_{par}(\mathbf{x}_S, \mathbf{x}_R) = t_0 + \mathbf{p}_R^T\mathbf{x}_R - \mathbf{p}_S^T\mathbf{x}_S - \mathbf{x}_S^T\mathbf{B}^{-1}\mathbf{x}_S^T + \frac{1}{2}\mathbf{x}_S\mathbf{B}^{-1}\mathbf{A}\mathbf{x}_S + \frac{1}{2}\mathbf{x}_R^T\mathbf{D}\mathbf{B}^{-1}\mathbf{x}_R, \quad (3.4)$$

where t_0 is the traveltime along the central ray. The parabolic traveltime approximation can be interpreted as the second-order expansion of the exact two-point traveltime between S and R, i.e., Equation 3.4 can be expressed in terms of traveltime derivatives of first and second order with respect to the displacement vectors $\mathbf{x}_S$ and $\mathbf{x}_R$. Introducing the 2×2 second-derivative matrices (see, e.g., Schleicher et al.,

1993).

$$N_S^R = \left(\frac{\partial^2 t}{\partial (x_S)_i \partial (x_S)_j} \right)_{i,j=1,2} = \mathbf{B}^{-1} \mathbf{A}, \qquad (3.5)$$

$$N_R^S = \left(\frac{\partial^2 t}{\partial (x_R)_i \partial (x_R)_j} \right)_{i,j=1,2} = \mathbf{D} \mathbf{B}^{-1}, \qquad (3.6)$$

$$N_{SR} = \left(\frac{\partial^2 t}{\partial (x_S)_i \partial (x_R)_j} \right)_{i,j=1,2} = \mathbf{B}^{-1}. \qquad (3.7)$$

Equation 3.4 can be rewritten in the form

$$t_{par}(\mathbf{x}_S, \mathbf{x}_R) = t_0 + \mathbf{p}_R^T \mathbf{x}_R - \mathbf{p}_S^T \mathbf{x}_S - \mathbf{x}_S^T N_{SR} \mathbf{x}_S^T + \frac{1}{2} \mathbf{x}_S N_S^R \mathbf{x}_S + \frac{1}{2} \mathbf{x}_R^T N_R^S \mathbf{x}_R \qquad (3.8)$$

Here, the subscript S and R means derivative with respect to shot and receiver coordinate while superscript stands for constant shot and receiver coordinate. Note that the matrices N_S^R and N_R^S are symmetric whereas N_{SR} is not. Although the Equation 3.8 is of fundamental importance, this parabolic travel time is not the best existing second-order approximation. It is generally known that in simply layered media, seismic near-vertical reflections are better approximated by hyperbolic rather than by parabolic traveltime curves (see, e.g., Ursin, 1982; Vanelle, 2002). Squaring Equation 3.8 and retaining only terms up to second order in $\mathbf{x}_S$ and $\mathbf{x}_R$ the hyperbolic traveltime approximation is given by

$$t_{hyp}^2(\mathbf{x}_S, \mathbf{x}_R) = \left(t_0 + \mathbf{p}_R^T \mathbf{x}_R - \mathbf{p}_S^T \mathbf{x}_S \right)^2 + t_0 \left(\mathbf{x}_S N_S^R \mathbf{x}_S - 2 \mathbf{x}_S^T N_{SR} \mathbf{x}_S^T + \mathbf{x}_R^T N_R^S \mathbf{x}_R \right). \qquad (3.9)$$

In the next section, I will show an example of the hyperbolic traveltime approximation, namely, common-reflection-surface stacking operator.

3.2 Common-reflection-surface stack

The general idea of the CRS technique is to describe a reflection event in the vicinity of a ZO sample by means of a second-order traveltime approximation given by Equation 3.9. The subsequent considerations are based on this hyperbolic traveltime representation given in midpoint and half-offset coordinates. If a source and a receiver with coordinates $\mathbf{x}_S$ and $\mathbf{x}_R$ are given, midpoint and half-offset coordinates are provided by the relations

$$\mathbf{m} = \frac{1}{2}(\mathbf{x}_S + \mathbf{x}_R), \quad \mathbf{h} = \frac{1}{2}(\mathbf{x}_S - \mathbf{x}_R) \tag{3.10}$$

For midpoint and half-offset displacement vectors Δm and Δh Equation s 3.10 can be extended to

$$\Delta \mathbf{m} = \mathbf{m} - \mathbf{x}_0 = \frac{1}{2}(\mathbf{x}_S + \mathbf{x}_R), \quad \Delta \mathbf{h} = \mathbf{h} - \mathbf{h}_0 = \frac{1}{2}(\mathbf{x}_S - \mathbf{x}_R) . \tag{3.11}$$

Here x_0 and h_0 refer to the central ray.

From the above equation source and receiver coordinates can be expressed as

$$\mathbf{x}_S = \Delta \mathbf{m} - \Delta \mathbf{h}, \quad \mathbf{x}_R = \Delta \mathbf{m} + \Delta \mathbf{h} \tag{3.12}$$

Substituting the expression for source and receiver coordinates in Equation 3.9 yields the finite-offset hyperbolic traveltime approximation in midpoint and half-offset coordinates (see, e.g., Spinner, 2007)

$$t(\mathbf{m},\mathbf{h})^2 = \left(t_0 + \Delta \mathbf{m}^T (\mathbf{p}_R - \mathbf{p}_S) + \Delta \mathbf{h}^T (\mathbf{p}_R - \mathbf{p}_S)\right)^2 + \tag{3.13}$$
$$t_0 \left(\Delta \mathbf{m}^T \mathbf{M}_M^H \Delta \mathbf{m} + 2\Delta \mathbf{m}^T \mathbf{M}_{MH} \Delta \mathbf{h} + \Delta \mathbf{h}^T \mathbf{M}_H^M \Delta \mathbf{h}\right),$$

where $t_0 = t(\mathbf{m}_0, \mathbf{h}_0)$ and

$$\mathbf{M}_M^H = \left(\frac{\partial^2 t}{\partial (m)_i \partial (m)_j}\right)_{i,j=1,2} = N_S^R + N_R^S + N_{SR} + N_{SR}^T, \tag{3.14}$$

$$\mathbf{M}_H^M = \left(\frac{\partial^2 t}{\partial (m)_i \partial (m)_j}\right)_{i,j=1,2} = N_S^R + N_R^S - N_{SR} - N_{SR}^T,$$

$$\mathbf{M}_{MH} = \left(\frac{\partial^2 t}{\partial h_i \partial h_j}\right)_{i,j=1,2} = -N_S^R + N_R^S + N_{SR} - N_{SR}^T.$$

In order to find the zero-offset hyperbolic traveltime approximation, one can consider only rays which are normal to the reflector in the so-called normal-incidence point (NIP). For these rays, the up- and down-going ray paths coincide, therefore $h_0 = 0$ and $\Delta h = h$. Consequently, the mixed second derivatives $\mathbf{M}_{MH}$ is zero. Moreover, the slowness at the receiver $\mathbf{p}_R$ equals the one of the source, $-\mathbf{p}_S$, and the following relation holds

$$\mathbf{p}_m = \left(\frac{\partial t}{\partial (m)_i}(\mathbf{x}_0, \mathbf{h}_0)\right) = \mathbf{p}_R - \mathbf{p}_S. \tag{3.15}$$

The zero-offset hyperbolic traveltime approximation in midpoint and half-offset coordinates is then

$$t(\mathbf{m}, \mathbf{h})^2 = \left(t_0 + 2\Delta\mathbf{m}^T \mathbf{p}_m\right)^2 + t_0 \left(\Delta\mathbf{m}^T \mathbf{M}_M^H \Delta\mathbf{m} + \mathbf{h}^T \mathbf{M}_H^M \mathbf{h}\right). \tag{3.16}$$

The stacking parameters $\mathbf{p}_m$, $\mathbf{M}_M^H$, and $\mathbf{M}_H^M$ contain information on the kinematics of the recorded wavefield. Physically, these parameters can be interpreted as attributes of two hypothetical wavefronts emerging at the measurement surface location m_0 (Hubral, 1983). The parameter $\mathbf{p}_m$ defines the horizontal component of the slowness vector of the zero-offset normal ray emerging at m_0. It is related to the emergence angle, β_0, and the azimuth of the emerging ray measured versus the x-axis, ϕ:

$$\mathbf{p}_m = \frac{1}{v_0}\left(\begin{array}{c} \cos\phi \sin\beta_0 \\ \sin\phi \cos\beta_0 \end{array}\right) \tag{3.17}$$

where v_0 is the near-surface velocity.

The parameter M_H^M is related to the curvature of a wavefront emerging at m_0 when

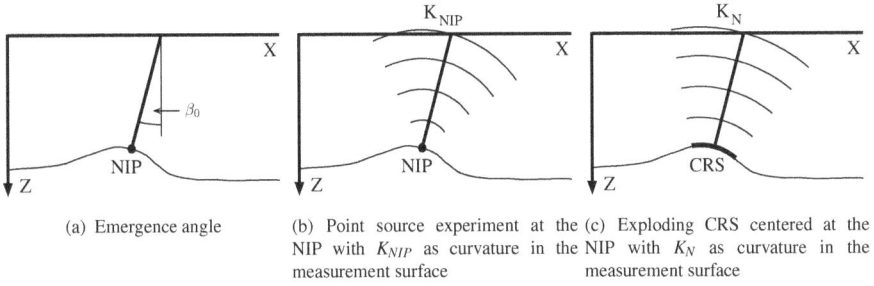

| (a) Emergence angle | (b) Point source experiment at the NIP with K_{NIP} as curvature in the measurement surface | (c) Exploding CRS centered at the NIP with K_N as curvature in the measurement surface |

Figure 3.2: Physical meaning of the CRS parameters, β_0, K_{NIP} and K_N, for a simple 2D example. These figures are kindly provided by Dr. Claudia Vanelle.

a point source is placed at the Normal-Incident-Point (NIP) of the reflector. The associated wave is called the NIP-wave. For the sake of simplicity, I will denote this parameter as $\mathbf{M}_{NIP}$. The expression for $\mathbf{M}_{NIP}$ reads as

$$\mathbf{M}_{NIP} = \frac{2}{v_0}\mathbf{H}\mathbf{K}_{NIP}\mathbf{H}^T, \qquad (3.18)$$

where $\mathbf{H}$ is the upper 2×2 sub-matrix of the transformation from ray-centered coordinate to the global coordinate system (Spinner, 2007) and $\mathbf{K}_{NIP}$ is the matrix of wavefront curvatures of the NIP-wave (Figure 3.2b).

The parameter M_M^H is related to the curvature of a wavefront emerging at m_0 from an exploding reflector element, the CRS element, centered at the NIP. Since all rays associated with this wave are locally normal to the reflector element in the subsurface, it is called the normal wave. For the sake of simplicity, I again will denote this parameter as $\mathbf{M}_N$. The expression for $\mathbf{M}_N$ reads as

$$\mathbf{M}_N = \frac{2}{v_0}\mathbf{H}\mathbf{K}_N\mathbf{H}^T, \qquad (3.19)$$

where $\mathbf{H}$ again stands for the transformation matrix and $\mathbf{K}_N$ is the wavefront curvature of the normal wave (Figure 3.2c).

The kinematic wavefield attributes may be used for a number of applications, including the calculation of geometrical spreading (Hubral, 1983), the determination of the approximated projected Fresnel zones (Mann, 2002), depth velocity model building with NIP-wave tomography (Duveneck, 2004), limited-aperture depth migration (Jäger, 2004), generalized Dix-type inversion (Müller, 2007), CRS-based time migration (Spinner, 2007), prestack data regularization and enhancement (Baykulov and Gajewski, 2009), or multiple suppression (Dümmong, 2010).

The 2D common-reflection-surface stack

In case of a 2D acquisition geometry, the seismic data is recorded along a single line which is usually defined as the x-axis of the global Cartesian coordinate system. This direction is commonly referred to as in-plane or inline direction whereas the y-coordinate defines the out-of-plane or cross-line direction. If the properties of the subsurface do not vary in the cross-line direction, all rays remain within the vertical observation plane defined by the acquisition line. In 2D, the midpoint and half-offset coordinates reduce to scalars and the traveltime approximation 3.16 can be expressed in terms of first and second traveltime derivatives:

$$t_{hyp}^2(\Delta m, h) = \left(t_0 + \frac{\partial t}{\partial m}\Delta m\right)^2 + t_0\left(\frac{1}{2}\frac{\partial^2 t}{\partial^2 m}\Delta m^2 + \frac{1}{2}\frac{\partial^2 t}{\partial^2 h}\Delta h^2\right). \qquad (3.20)$$

In the same way as in the 3D case, the traveltime derivatives can be related to physical properties of the subsurface using Equation s 3.17, 3.18, and 3.19. Due to the invariance of the model in y-direction, the azimuth ϕ equals zero. The first derivative of the traveltime with respect to the midpoint reduces to

$$\frac{\partial t}{\partial m} = \frac{2\sin\beta_0}{v_0} \qquad (3.21)$$

and second derivatives of the traveltime with respect to the midpoint and offset coordinate are given by

$$\frac{\partial^2 t}{\partial^2 m} = \frac{2\cos\beta_0}{v_0}K_N \quad \text{and} \qquad (3.22)$$

$$\frac{\partial^2 t}{\partial^2 h} = \frac{2\cos\beta_0}{v_0}K_{NIP} \qquad (3.23)$$

40

With Equation 3.21, 3.22, and 3.23, the final form of the 2D CRS operator yields

$$t_{hyp}^2(\Delta m, h) = \left(t_0 + \frac{2 \sin \beta_0}{v_0} \Delta m \right)^2 + \frac{2 t_0 \cos^2 \beta_0}{v_0} \left(K_N \Delta m^2 + K_{NIP} \Delta h^2 \right) \qquad (3.24)$$

The main advantage of the CRS operator is the use of an entire stacking surface in the time-midpoint-half-offset space while NMO approach uses a trajectory in time-half-offset plane. This implies that the considered reflection events are continuous over several neighboring midpoint gathers. This overcomes the problem that a CMP gather may contain information from more than one reflection point in depth, i.e., the CRS operator inherently accounts for the reflection point dispersal, i.e., the main part of the CRS trajectory, which belongs to a specific ZO sample, lies within the corresponding stacking surface. In Figure 3.3 the CRS stacking operator is depicted for a simple 2D model.

In this section, I have described a multiparameter stacking operator, the CRS operator, based on the hyperbolic traveltime expression of reflections. Although the stacked section provided by the CRS method exhibits an improved S/N, it still may be contaminated by triplications and diffractions which may lead to conflicting dip situations and thus complicate the proper determination of kinematic wavefield parameters. To properly determine kinematic wavefield attributes the data may be transformed into another domain where these undesired effects vanish. In this context, it is reasonable to transform data in time-magrated domain. One of the suitable techniques to produce time-migrated data is the CSP data mapping (see Chapter 2 for detailed description of the CSP data mapping). Since the CSP data represent 'partly' time-migrated data which should be focused at zero-offset they are very suitable to estimate kinematic wavefield parameters in the time-migrated domain. In the next section, I will present a multiparameter stacking technique based on the hyperbolic traveltime expression of time-migrated reflections.

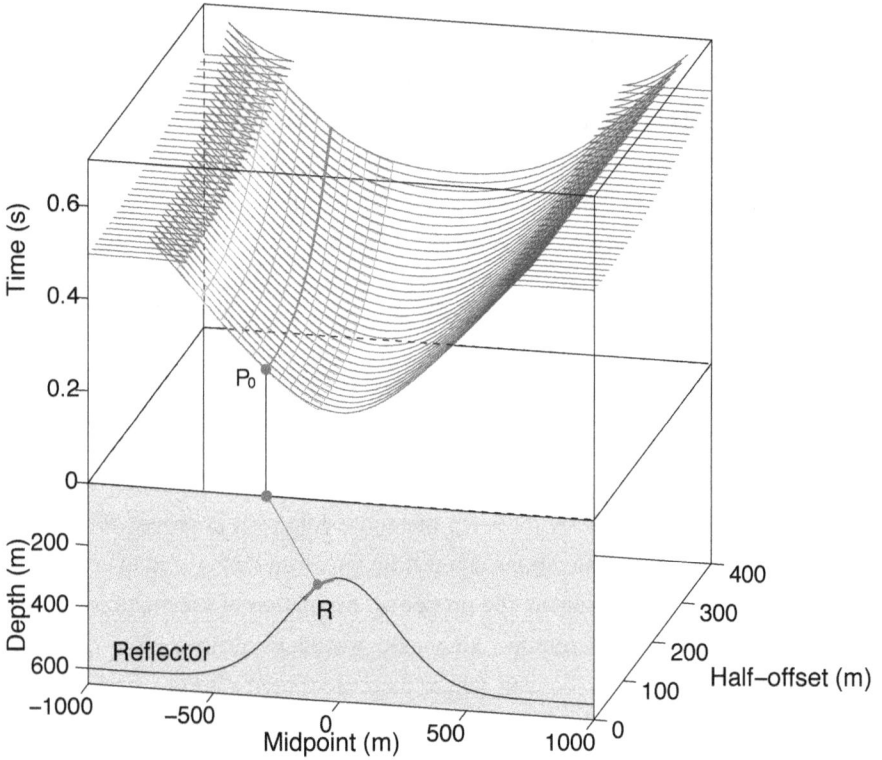

Figure 3.3: CRS stacking operator and CRP trajectory in the 2D time-midpoint-offset volume. The blue curves in the upper part of the pictures represent reflection traveltime curves for fixed source-receiver offsets for the dome-like reflector in the lower part. The green lines indicate the CRS stacking operator for the ZO sample P_0 which approximates the reflection response of the red reflector segment around the Common-Reflection-Point (here set up by means of neighboring CRP trajectories).

42

3.3 Common-migrated-reflector-element stack

In this section, I am going to present a new multiparameter stacking technique, the common-migrated-reflector-element (CMRE) stack, which is described by three stacking parameters. These parameters are like CRS parameters. For CMP data they apply to the NIP-ray. For CSP data they apply to the image ray. They represent spatial traveltime derivatives of the first- and second-order for the time-migrated reflections. The first-order derivative contains information on the time-migrated reflector. The second-order derivative with respect to the scatter point displacement contains information on the curvature of the time-migrated reflector (Tygel et al., 2010). The second-order derivative with respect to offset contains information on the wavefront curvature of the image ray (Dell et al., 2010). The CMRE stack applied to the CSP gathers generates the time-migrated image. The CMRE stack comprises three automatic searches and represents a high-density velocity analysis. The first search determines best-fit stacking parameter in offset direction. The remaining searches determine best-fit stacking parameters in midpoint direction. Thus, the CMRE stack provides an updated time-migration velocity model with improved vertical and lateral resolution. Also the CMRE stack accounts for the neighboring CSPs so that more traces can be involved in the stacking resulting in an enhanced signal-to-noise ratio of the time-migrated section.

Similarly to the CRS approach, we can expand the traveltime of time-migrated reflections as a Taylor series up to second order and then square it. However, instead of expanding the traveltime for the normal ray, we now expand the traveltime in the vicinity of the image ray. The image ray intersects the reflector at the Image-Incident-Point (IIP) (Tygel et al., 2009). Furthermore, we consider a continuous reflector segment centered at the IIP. This reflector segment we call a migrated reflector element (see Figure 3.4). The paraxial image rays, which are reflected on the migrated reflector element, build its seismic response in the prestack domain of time-migrated reflections. The new operator approximates the response of a whole migrated-reflector element and focuses time-migrated reflections at zero-offset.

The traveltime surface of the migrated reflector element is given in terms of image

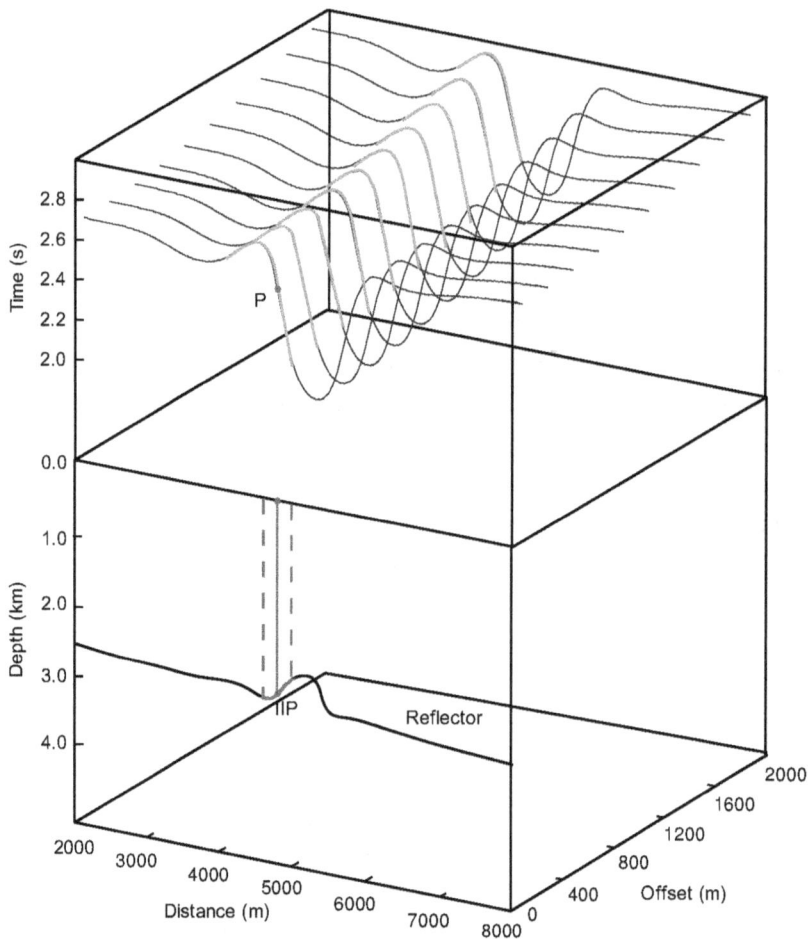

Figure 3.4: Illustration of CMRE stacking operator in the 2D time-midpoint-offset volume. The blue curves in the upper part of the picture represent traveltimes of time-migrated reflections for fixed scatterpoint offsets for the dome-like reflector in the lower part. The green lines indicate the CMRE stacking operator for the ZO sample P which approximates the response of the red reflector element around the Image-Incident-Point. Please note, that there is no folding in the time-migrated traveltimes.

point coordinates $(\mathbf{x}, \mathbf{h}_s, T^M)$ and reads as

$$T^M(\mathbf{x}, \mathbf{h}_s)^2 = \left(T_0^M + \frac{\partial T^M}{\partial x_i} \mathbf{x}^T \right)^2 + 2T_0^M \left(\mathbf{x}^T \frac{\partial^2 T^M}{\partial x_i \partial x_j} \mathbf{x} + \mathbf{h}_s^T \frac{\partial^2 T^M}{\partial (h_s)_i \partial (h_s)_j} \mathbf{h}_s \right),$$
(3.25)

where $\mathbf{x}$ now specifies the scatterpoint displacement vector with respect to the considered CSP, $\mathbf{h}_s$ now is the scatterpoint offset vector which is a half of the distances from the scatterpoint location to the collocated source and receiver, and T_0^M is twice the traveltime along the central image ray from the reference point to the IIP. The first-order derivatives $\partial T^M / \partial x_i$ are related to the dip $\mathbf{p}^M$ of the time-migrated reflector. The second-order derivatives with respect to the scatterpoint displacement $\partial^2 T^M / \partial (x)_i \partial (x)_j$ is related to the curvature of the time-migrated reflector element, $\mathbf{M}_{xx}^M$ (Tygel et al., 2010). The second-order derivative with respect to the scatterpoint offset $\partial^2 T^M / \partial (h_s)_i \partial (h_s)_j$ are related to the wavefront curvature of the image ray, $\mathbf{M}_{h_s h_s}^M$ (Dell et al., 2010).

Substituting the formulas for the derivatives, i.e.,

$$\mathbf{p}^M = \frac{1}{2} \left(\frac{\partial T^M}{\partial (x)_i} \right), \mathbf{M}_{xx}^M = \frac{1}{2} \left(\frac{\partial^2 T^M}{\partial x_i \partial x_j} \right) \mathbf{M}_{h_s h_s}^M = \frac{1}{2} \left(\frac{\partial^2 T^M}{\partial (h_s)_i \partial (h_s)_j} \right)$$

in Equation 3.25 yields

$$T^M(\mathbf{x}, \mathbf{h}_s)^2 = \left(T_0^M + 2\mathbf{p}^M \mathbf{x}^T \right)^2 + 2T_0^M \left(\mathbf{x}^T \mathbf{M}_{xx}^M \mathbf{x} + \mathbf{h}_s^T \mathbf{M}_{h_s h_s}^M \mathbf{h}_s \right).$$
(3.26)

The above expression looks formally similar to the conventional CRS operator. However, CMRE-stacking parameters are model-based while the CRS stacking parameter are surface-related.

To allow for better interpretation of the stacking parameters I will split the CMRE operator. For the zero-offset the CMRE operator simplifies to the following expression

$$T^M(\mathbf{x}, \mathbf{h}_s)^2 = \left(T_0^M + 2\mathbf{p}^M \mathbf{x}^T \right)^2 + 2T_0^M \mathbf{x}^T \mathbf{M}_{xx}^M \mathbf{x}.$$
(3.27)

Also I make use of ray-propagator matrices. In this way, one can introduce the 4×4 surface-to-surface propagator matrix of the central downgoing image ray (Tygel et al., 2009)

$$\underline{\mathbf{T}}(R,S) = \begin{pmatrix} \mathbf{A} & \mathbf{B} \\ \mathbf{C} & \mathbf{D}, \end{pmatrix}$$

which connects the anterior surface to the posterior surface (Figure 3.1). The initial and end points of the central image ray are specified by the coordinates S_0 and R_0, respectively. With the above considerations, the one-way traveltime of a paraxial image ray $S_0(R_0)$ can be expressed as a function of its coordinates at the anterior surface ($\mathbf{x}_S$) and the posterior surface ($\mathbf{x}_R$), respectively, by a second-order approximation which reads as

$$t_{par}(\mathbf{x}_S, \mathbf{x}_R) = t_0 + \mathbf{p}_R^T \mathbf{x}_R - \mathbf{p}_S^T \mathbf{x}_S - \mathbf{x}_S^T \mathbf{B}^{-1} \mathbf{x}_S^T + \frac{1}{2} \mathbf{x}_S \mathbf{B}^{-1} \mathbf{A} \mathbf{x}_S + \frac{1}{2} \mathbf{x}_R^T \mathbf{D} \mathbf{B}^{-1} \mathbf{x}_R.$$
(3.28)

The quantity $\mathbf{p}^T(R_0)$ is the projection of the slowness vector of the central ray on the tangent plane to the posterior surface at its end point. Taking into account that for an image ray the slowness projection, $\mathbf{p}^T(S_0)$, on the tangent plane to the anterior surface at its initial point vanishes, and the case of zero-offset ray, the expression above simplifies to

$$t_{par}(\mathbf{x}_S, \mathbf{x}_R) = t_0 + \mathbf{p}_R^T \mathbf{x}_R - \mathbf{x}_S^T \mathbf{B}^{-1} \mathbf{x}_S^T + \frac{1}{2} \mathbf{x}_S \mathbf{B}^{-1} \mathbf{A} \mathbf{x}_S + \frac{1}{2} \mathbf{x}_R^T \mathbf{D} \mathbf{B}^{-1} \mathbf{x}_R.$$
(3.29)

The coordinates S_0 and R_0, which refer to the initial and endpoints of a specified paraxial image ray, are connected. To find the connection, we investigate the case and recall the properties of the propagator matrix to write

$$\Delta \mathbf{x}_R = \mathbf{A} \mathbf{x}_S + \mathbf{B} \Delta \mathbf{p}_S = \mathbf{A} \mathbf{x}_S.$$
(3.30)

Expressing $\Delta \mathbf{x}_R$ as a second-order expansion of $\Delta \mathbf{x}_S$ provides (see Tygel et al., 2010):

$$\Delta x_R^k = \frac{\partial x_R^k}{\partial x_S^j} \Delta x_S^j + \frac{1}{2} \frac{\partial^2 x_R^k}{\partial x_S^i \partial x_S^j} \Delta x_S^i \Delta x_S^j,$$
(3.31)

where upper indices are the summation indices. One can recognize that the first derivative in this equation equals the matrix element A_{kj}, while the second derivative equals $\partial A_{kj}/\partial m_i$ (Tygel et al., 2010). Making use of the relation

$$\mathbf{A}^T \mathbf{D} - \mathbf{C}^T \mathbf{B} = \mathbf{I} \tag{3.32}$$

results in

$$t_{par}(\mathbf{x}_S, \mathbf{x}_R) = t_0 + \mathbf{x}_S^T \mathbf{A}^T \mathbf{p}_R + \frac{1}{2} \mathbf{x}_S \left(\mathbf{E} + \mathbf{C}^T \mathbf{A} \right) \mathbf{x}_S, \tag{3.33}$$

where the elements of matrix $\mathbf{E}$ are given by

$$E_{ij} = \frac{\partial^2 x_R^k}{\partial x_S^i \partial x_S^j} p_R^k = \frac{\partial A_{kj}}{\partial m_i} p_R^k. \tag{3.34}$$

Squaring Equation 3.29 and retaining only terms up to second order in $\mathbf{x}_S$ and $\mathbf{x}_R$ the hyperbolic traveltime approximation is the given by (Tygel et al., 2010)

$$t_{hyp}^2(\mathbf{x}_S, \mathbf{x}_R) = \left(t_0 + \mathbf{x}_S^T \mathbf{A}^T \mathbf{p}_R \right)^2 + t_0 \mathbf{x}_S^T \left(\mathbf{E} + \mathbf{C}^T \mathbf{A} \right) \mathbf{x}_S^T. \tag{3.35}$$

Considering the observed two-way traveltime parameters are consistent with the one-way traveltime parameters simulated by paraxial image rays (Tygel et al., 2010), i.e., comparing Equations 3.27 and 3.35, yields

$$T_0^M = 2t(S_0; R_0), \tag{3.36}$$

$$\mathbf{p}^M = 2\mathbf{A}^T \mathbf{p}_R, \tag{3.37}$$

$$M_{xx}^M = 2(\mathbf{E} + \mathbf{C}^T \mathbf{A}). \tag{3.38}$$

One can conclude that the parameter M_{xx}^M can be realted to the curvature of the unmigrated reflector and parameter $\mathbf{p}^M$ to the reflector dip.

Now I investigate the case when $\mathbf{x} = 0$ in order to find a physical interpretation of parameter $\mathbf{M}_{h_s h_s}^M$. The expression for the CRME operator for time-migrated reflections given by Equation 3.26 becomes

$$t(\mathbf{h}_s)^2 = t_0^2 + 2t_0 \, \mathbf{h}_s^T \mathbf{M}_{h_s h_s}^M \mathbf{h}_s. \tag{3.39}$$

Comparing this Equation with the traveltime approximation for paraxial rays given

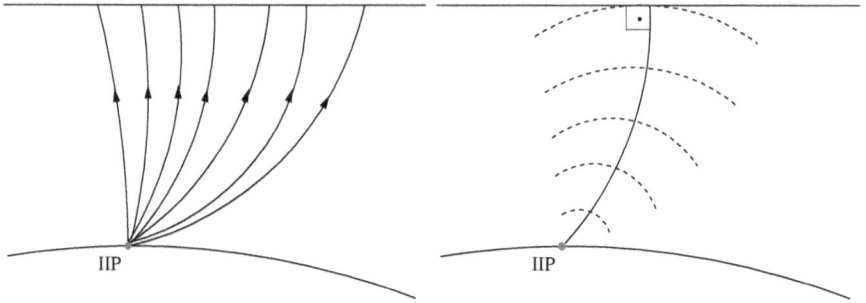

Figure 3.5: The ray trajectories associated with a hypothetical wave due to a point source at the IIP (IIP wave). Geometrically, the CSP ray segments build the IIP wave (a). In a consistent velocity model, IIP waves focus at the IIP at zero traveltime, when they are propagated back in the time-migrated domain (b).

by Equation 4.18, we see that the matrix $\mathbf{M}^M_{h_s h_s}$ is linked to the wavefront curvature of the IIP-wave, i.e., the wavefront corresponding to the image ray.

Similar to the NIP-wave experiment (Chernyak and Gritsenko, 1979; Hubral and Krey, 1980), we can consider an IIP-wave experiment. For the sake of simplicity, I consider the 2D case only. The experiment can be carried out by placing a point source in the IIP of the central image ray. The wave propagates along a central image ray to the measurement surface (Figure 3.5). Note that for an image ray the slowness projection on the tangent plane to the emergence location at its initial point vanishes. Therefore, at the emergence location $(x = 0, h_s)$ the central image ray is determined in terms of the curvature of the IIP-wave, i.e., K_{IIP}.

The expression given in Equation 3.26 represents the CMRE operator approximates the traveltime of a time-migrated reflector element centered around the IIP. Application of the CMRE operator to CSP data focuses scattered energy to zero-offset and completes prestack time migration in this workflow. The first step in the CMRE stack is the automatic CSP stack which is similar to an automatic CMP stack. The automatic CSP stack provides the stacking parameters. The parameter searches are carried out individually by using a coherence measure which is calculated for

48

a number of test parameters. The parameter leading to the highest coherence value and stack amplitude, is considered to be the searched-for parameter. The searches are applied without any user interaction only guided by a coherence section. To increase the reliability of the parameters, the search process is finalized by a subsequent local simultaneous optimization of all parameters using the previously determined initial values (Hertweck et al., 2007) and an event-consistent smoothing (Mann and Duveneck, 2004). The event-consistent smoothing is based on a combination of median filtering and averaging.

3.4 Synthetic data example

In this section I apply the formally presented traveltime approximations to a synthetic example. Figure 3.6a displays a simple synthetic model with an anticline in the middle. The model consists of 3 layers. The velocity in the first layer is 1500 m/s, in the second layer 2300 m/s, and in the third layer 2700 m/s. The synthetic seismograms were generated by a ray tracer package kindly provided by NORSAR Innovation AS. I used a Ricker-wavelet with a prevailing frequency of 25 Hz. The sampling interval is 2 ms and the used recording time is 2.4 s. Gaussian noise was added to the prestack data. The signal-to-noise ratio is 20. CMP gathers are displayed in Figure 3.6b. The CRS stacked section displays triplications caused by the anticline structure (see Figure 3.7a).

Firstly, the automatic CMP stack was applied to prestack gathers to determine stacking velocities. The stacking velocities then served as initial time-migration velocities to generate CSP gathers (see Figure 3.6c). Afterwards, the CMRE stack was applied to CSP gathers to extract curvatures of the IIP-waves. Figure 3.7 shows the CRS and CMRE stacked sections. Triplications are unfolded in the CMRE stack section while they are present in the CRS stacked section. The high focusing of the triplications indicates that time-migration velocities were correctly determined. Also, the CMRE stack provides a higher coherence section (Figure 3.8).

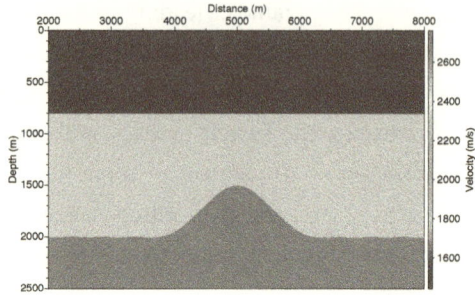

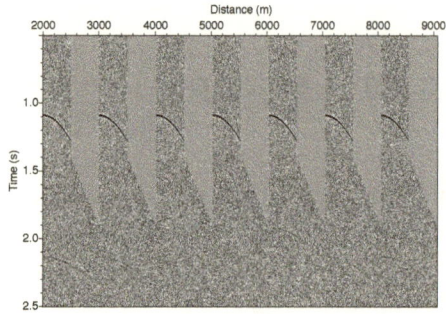

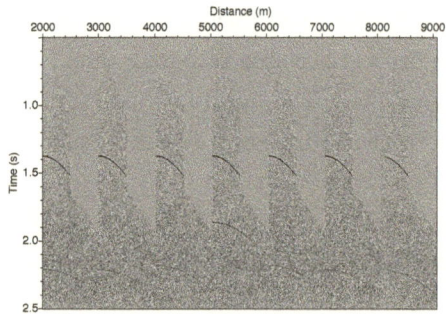

Figure 3.6: A simple synthetic model with an anticline in the middle. The model consists of 3 layers. The velocity in the first layer is 1500 m/s, in the second layer 2300 m/s, in the third layer 2700 m/s. (b) CMP gathers. Note that because of higher noise level in the data the second reflections is not observerable in the CMP gather. (c) CSP gathers. Because of the improved S/N the second reflection is now visible.

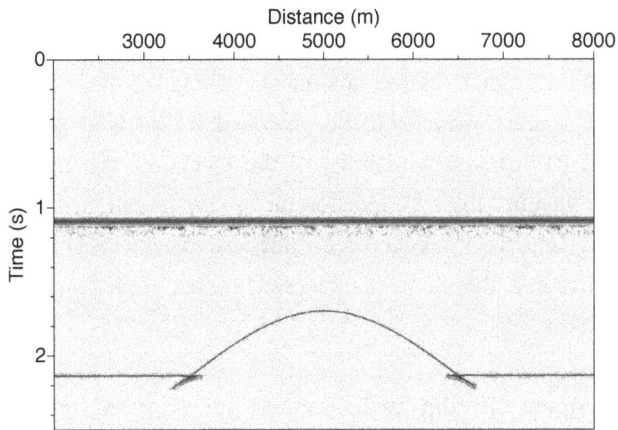

(a)

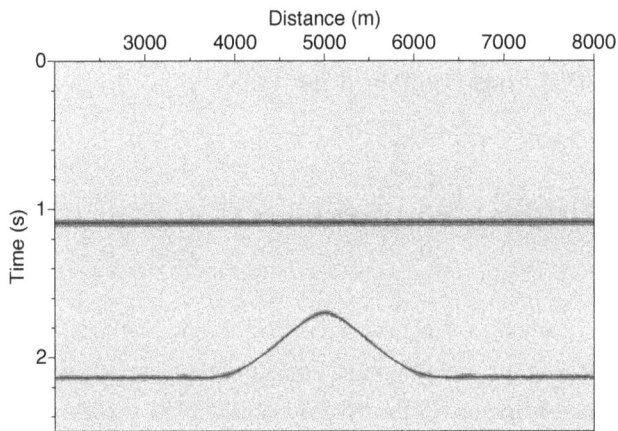

(b)

Figure 3.7: Stacks of the simple synthetic model with an anticline (see Figure 3.6a).
The CRS-stacked section (a) displays triplications caused by the anticline
structure. The CMRE-stacked section (b) shows the anticline structure
and no triplications. Both sections were weighted with the semblance
value of 0.3.

51

Especially for the top of the anticline, we observe a significant rise of the coherence value in comparison to the coherence section provided by the CRS stack. The coherence value is crucial for the automatic picking of data vector components in an inversion process which will be proposed in the next section. The high coherence value confirms the reliability of the kinematic wavefield attributes and allows to prove whether the pick location under consideration is actually part of a reflection event. Also, the low coherence value lead to a decrease of the searched-for coherence maxima and, therefore, an increased number of picks which causes higher computational costs.

Figure 3.9 illustrates the dip angle sections for migrated (a) and unmigrated reflections (b). In the unmigrated dip section, we observe conflicting dips when triplications are close to the reflection while no conflicting dips are present in the migrated dip section. The same kinematic wavefield attributes behavior is observable in the R_N and the R_{MRE} sections (Figure 3.10). In the R_N section, we observe the ambiguity of the wavefield attribute determination close to triplications. In contrast, the determination of R_{MRE} results in unique values.

3.5 Concluding remarks

In this chapter, I among other have presented a new method to approximate the reflection response in the time-migrated domain. The new approximation is referred to as the Common-Migrated-Reflection-Element (CMRE) stack. The CMRE stack represents a multiparameter stacking technique based on the Taylor expansion of time-migrated traveltimes in the vicinity of the central image ray. To generate time-migrated reflections I applied the CSP data mapping to prestack data. The application of the CMRE stack to prestack-time migrated data provides a time-migrated section. The CMRE stack belongs to high-density velocity analysis tools with a minimal user interaction. Migration velocities estimated during the CMRE stack have a high horizontal and vertical resolution. Also the automatic CMRE stack considers neighboring CSP gathers, hence, an increased number of traces is stacked. Due to the

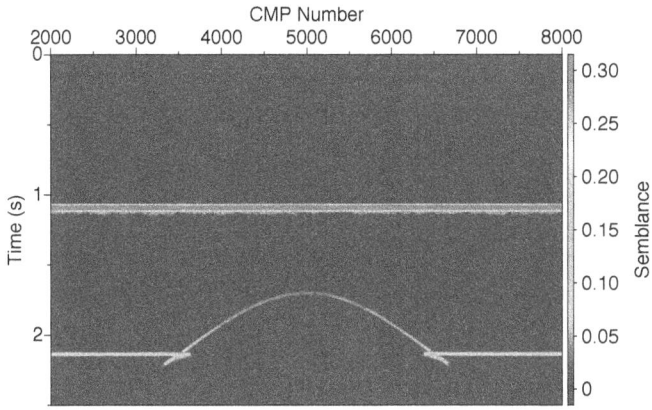

(a)

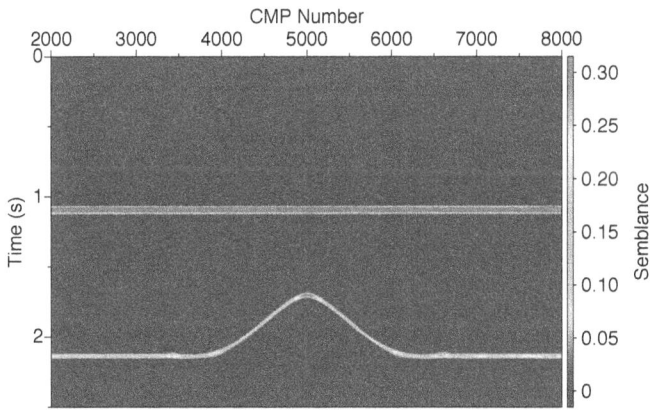

(b)

Figure 3.8: Coherency sections obtained by the CRS stack (a) and the CMRE stack (b). In the latter one, we observe higher coherence values for the second reflection. The high coherence value is crucial for picking of events in tomographic methods.

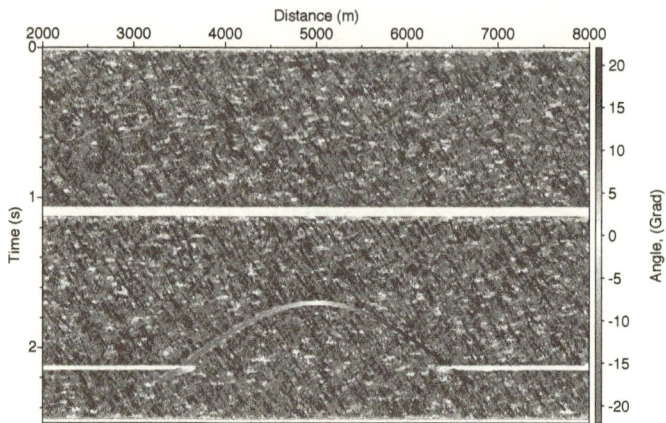

(a)

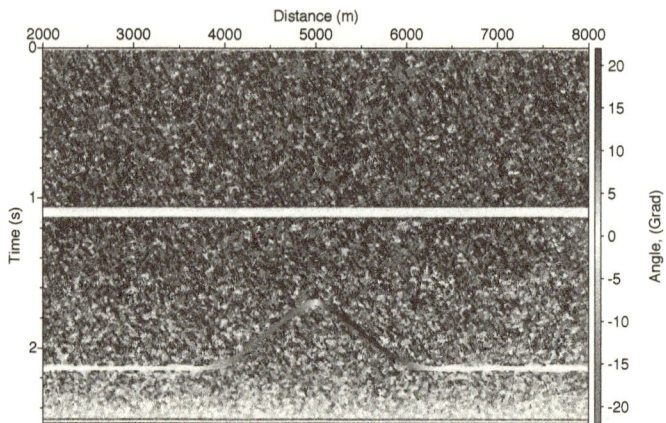

(b)

Figure 3.9: Angle sections. In the normal-angle section, we observe conflicting dips when triplications are close to the reflection (a) while no conflicting dips are present in the time-migrated dip section (b).

54

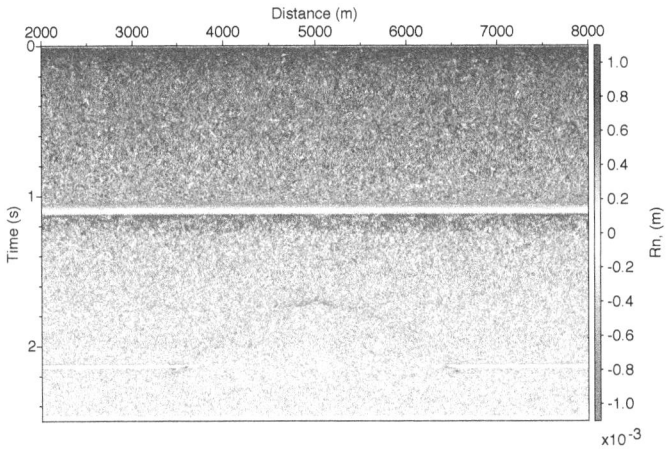

(a)

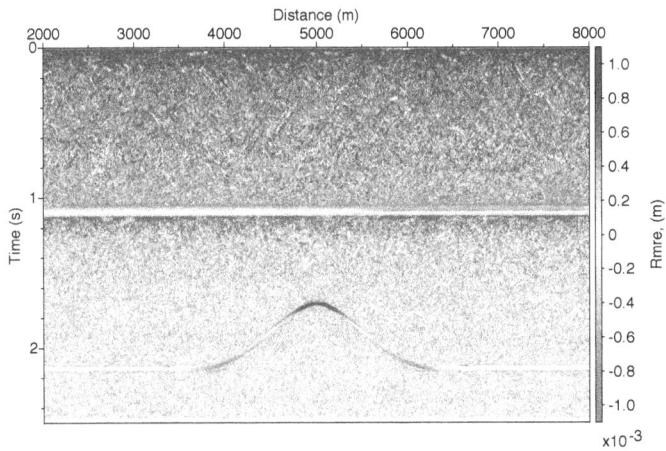

(b)

Figure 3.10: R_N section (a) and R_{MRE} section (b). In the R_N section, we observe the ambiguity of the wavefield attribute determination close to triplications. In the R_{MRE} section the determination of the wavefield attribute is unique.

55

improved time-migration velocities and stacking of more traces, the CMRE stack of CSP data provides eventually an enhanced and highly-focused time-migrated image in an automated manner.

4 Image ray tomography

The construction of velocity models is an important task for seismic depth imaging. Over the past decades many techniques have been developed based upon different approaches, e.g., methods based on migration velocity analysis or traveltime inversion. The main advantage of the reflection tomography based on traveltime inversion is that only locally coherent events are required to reconstruct the velocity distribution and no repetitive depth migration processes are involved. Reflection-tomography methods use kinematic information of traveltimes, such as dips in stereotomography (see, e.g., Billette and Lambaré, 1998) or wavefront curvatures and emergence angles in NIP wave tomography (see, e.g., Duveneck, 2004). However, the kinematic attributes are extracted either from prestack or CMP stacked data that are usually contaminated by diffractions and triplications. The reliable determination of kinematic attributes of reflections close to triplications and diffractions is very complicated. The potential biasing of attributes affects the inversion process resulting very often in erroneous velocity models and, subsequently, smeared depth images. To overcome these limitations, we propose to extract the kinematic attributes in the time-migrated domain. Time-migrated reflections represent 'purer' reflection data, since diffractions are collapsed and triplications unfolded. Therefore, kinematic attributes estimated from the time-migrated reflections domain are expected to bear a higher quality and easier picking.

The time-migrated reflections are related to image rays in a similar way as normal incidence rays relate to primary reflections. The time-migrated primary reflections can be obtained by tracing image rays vertically down from the surface to the desired reflector in depth (Hubral, 1977). The image ray is normal to the acquisition surface and, when propagating down, intersects the reflector at the Image Incident Point (IIP)

(Tygel et al., 2009). Due to the increase in reflector resolution in the time domain, it is preferable to use image rays for the velocity model building (Cameron et al., 2007; Iversen and Tygel, 2008). The concept of image rays has been limited to the poststack domain up to now. The kinematic and dynamic attributes of image rays are computed using ray tracing which is very sensitive to the initial time-migration velocity model. Moreover, the time-migrated velocities are frequently transformed in time-interval velocities by Dix inversion, which further contributes to distortion of the time-migration velocities. The above mentioned aspects make the velocity model building based on the poststack image-ray tracing inefficient in many cases.

I propose to use the wavefront curvatures of the image rays to construct velocity models in a tomographic manner. We assume also that curvature parameters of the image rays are already available. Furthermore, we describe the propagation of the image rays in terms of a hypothetical IIP-wave. The IIP-wave is a wave which propagates in a hypothetical homogeneous model when a point source explodes at the IIP. The velocity model is said to be consistent with the data if wavefront curvatures of the IIP-waves after propagating into the subsurface become vanishingly small at all considered IIP's. Since the phase slownesses of image rays are normal to the acquisition surface, we only have to worry about their wavefront curvatures. The model parameters are calculated by dynamic ray-tracing along central image rays. As a common practice in tomographic processes, the non-linear inversion problem is solved iteratively by using the least-squares solutions to locally linearized problems. The required model parameter (Fréchet) derivatives for the tomographic matrix are calculated using ray perturbation theory.

4.1 NIP-wave tomographic inversion

Before formulating the image-ray tomography problem, I will briefly review general principles of the NIP-wave tomography. The NIP-wave tomography represents an inversion method based on the kinematic of reflection traveltimes. The basic idea is to describe the CMP reflection traveltimes in terms of the propagation of a hypothetical NIP-wave. This representation requires the NIP-wave theorem to be valid (Chernyak and Gritsenko, 1979; Hubral, 1983). The theorem states that the CMP specular reflection traveltime along the ray is, up to second-order in half-offset, equal to the traveltime along a non-specular ray from a source via the NIP to a receiver. That means that each paraxial ray of a CMP ray family can be viewed as passing through the common NIP instead of the offset-dependent real reflection point (Schleicher et al., 2007). The theorem is illustrated in Figure 4.1. Traveltimes along specular rays connecting sources and receivers on the measurement surface and reflected at a reflector segment (black lines) in the subsurface coincide with traveltimes along paraxial rays in the vicinity of a central normal ray reflected at the NIP (blue lines). These paraxial rays represent ray trajectories associated with a hypothetical emerging wave due to a point source at the NIP (Figure 4.1b). If reflection traveltimes in the seismic data can be interpreted in terms of the NIP wave, the focusing of this wave at zero traveltime at the NIP is equivalent to focusing CMP-traveltimes at real offset-dependent reflection points when propagated back into the subsurface. Thus, a model in which all NIP waves focus at zero traveltime is consistent with the data. Although the equivalence of traveltimes guaranteed by the NIP-wave theorem is valid only in a certain region centered at the CMP, i.e., small offsets, it can be generalized to an extended NIP-wave theorem valid for larger offsets (Schleicher et al., 2007).

Mathematically, the NIP-wave theorem represents the following equality

$$t_{CMP}(h) = t_0 + \frac{1}{2}h^T\frac{\partial^2 t}{\partial h^2}h = t_0 + M_{NIP}^{(x_0)}h^2 = t_{NIP}(h,x_0) \qquad (4.1)$$

where h is the half-offset, x_0 is the central midpoint coordinate, $M_{NIP}^{(x_0)}$ is determined as $M_{NIP}^{(x_0)} = \frac{2}{v_0}K_{NIP}$, where K_{NIP} is the curvature of the NIP wave emerging at x_0 and v_0 is the near-surface velocity. We will formulate the inversion problem for the 2-D case, however, an extension to 3-D is straightforward. The data vector is unambiguously determined by the normal ray traveltime t_0, the first-order spatial traveltime derivative $p(x)$, and the second-order spatial traveltime derivative M_{NIP} at the respective normal ray emergence location x_0. These parameters can be extracted from the seismic prestack data by, e.g., applying conventional stacking velocity

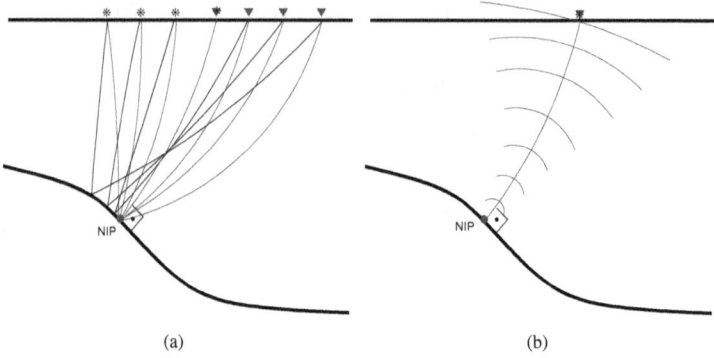

Figure 4.1: Illustration of the NIP theorem. (a) The traveltime along specular rays connecting sources and receivers on the measurement surface for a central midpoint reflected at a reflector segment (black lines) in the subsurface coincide with traveltimes along paraxial rays in the vicinity of a central normal ray reflected at the NIP. (b) The paraxial-ray trajectories are associated with a hypothetical wave due to a point source at the NIP. This wave is referred to as NIP wave.

analysis and a local zero-offset dip search.

The data vector is then given as

$$\mathbf{d} = (t_0, M_{NIP}, p, x)_i, \ i = 1, ..., n \tag{4.2}$$

where n is the number of picked data points. Each of these data points is associated with a NIP in the subsurface, characterized by its spatial location $(x, z)^{(NIP)}$.

The 2-D velocity model is described by two-dimensional B-splines

$$v(x, z) = \sum_{j=1}^{n_x} \sum_{k=1}^{n_z} m_{jk} \beta_j(x) \beta_k(z) \tag{4.3}$$

where m_{jk} are B-spline coefficients representing the velocity model on a rectangular grid, n_x and n_z are the numbers of grid points in the horizontal and vertical directions.

The model vector contains the calculated position of the NIP in the subsurface, $(x,z)^{(NIP)}$, computed values of velocity coefficients, m_{jk}, $j = 1, ..., n_x$, $k = 1, ..., n_z$, and calculated wavefront curvatures of the normal rays and its horizontal slowness projection for all data points

$$\mathbf{d}_{mod} = (t_0, M, p, x)_i^{mod}, \ i = 1, ..., n. \tag{4.4}$$

The model parameter $M = M_{NIP}$ can be calculated for the point-source initial condition as (see, e.g., in Červený, 2001; Popov, 2002)

$$M = P_2 Q_2^{-1}. \tag{4.5}$$

where P_2 and Q_2 are matrices represent a solution of the dynamic ray-tracing system for point-source initial conditions (see equation 4.9). The inverse problem to be solved is to minimize the energy function between the data vector $\mathbf{d}$ and the corresponding model vector $\mathbf{d}_{mod}$.

The generalized algorithm for NIP-wave tomography contains the following steps (see, e.g., in Duveneck, 2004):

1. The initial velocity model is set up using an initial velocity gradient.

2. The data points $\mathbf{d}$ are traced into the model until the traveltime t_0 is encountered. This provides NIP locations and normal ray directions in depth.

3. Dynamic ray tracing is used to forward model the vector $\mathbf{d}_{mod}$ and the Fréchet derivatives $\mathbf{F}$ used for the model update.

4. The energy function is evaluated from $\mathbf{d}$ and $\mathbf{d}_{mod}$.

5. An updated model is calculated. Within this model ray tracing is performed to obtain a new vector $\mathbf{d}_{mod}$.

6. The energy function is evaluated with the newly obtained $\mathbf{d}_{mod}$.

7. If the energy function has increased, the model update is rejected and the last two steps are repeated until the energy function decreases.

8. If the energy function has decreased, the updated model is used in the next iteration starting with step 3. The process is repeated until a maximum number of iterations is reached or the energy function falls below a threshold. In this case, the updated model is assumed to be the final inversion result.

4.2 Review of the image ray concept

In this section, I will review Hubral's image-ray concept. To determine the depth-migrated position of the focused energy, (Hubral, 1977) introduced the concept of the image ray. The image ray connects a depth point with the surface position of its image, i.e., it is the propagation path of the energy in the signal to the apex position of the migration operator. As diffraction surfaces have zero time-slope at their apex positions, all image rays hit the acquisition surface vertically (Figure 4.2). The image rays are naturally vertical only in media with constant velocity. In media with lateral velocity variations they propagate in a similar way as normal incident rays. The image rays relate to migrated reflections in a similar way as normal incidence rays relate to stacked primary reflections. Migrated primary reflections can be obtained by tracing rays vertically down from the free surface to the desired reflector at depth (Hubral, 1977). Thus, kinematic attributes of the image rays can be used to reconstruct velocity models similar to the kinematic attributes of the normal rays.

Similar to the normal rays, the kinematic ray tracing for image rays in 3D media with a smooth velocity can be formulated by the ordinary differential equations (see, e.g., Červený, 2001).

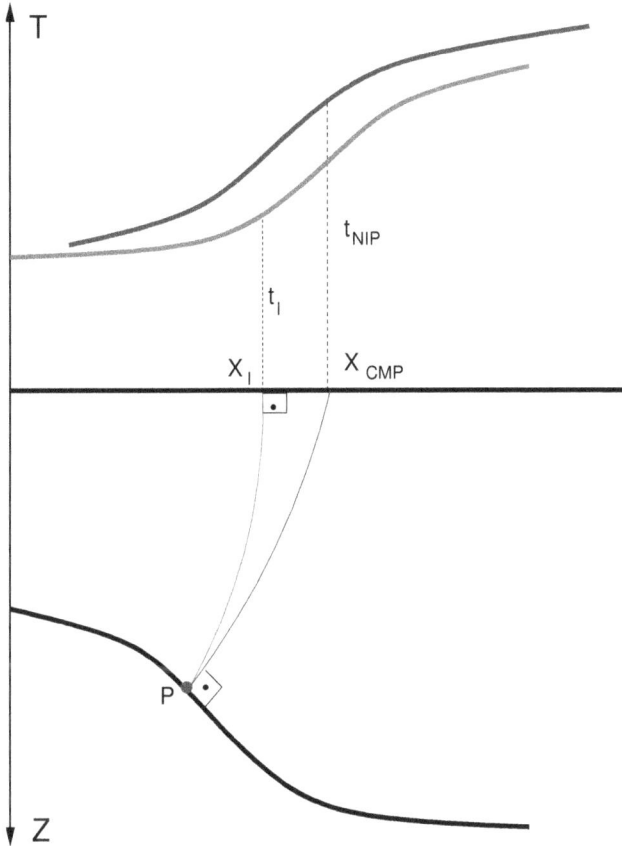

Figure 4.2: 2-D subsurface iso-velocity layer model of a reflector and its reflection response (depicted in blue) and time-migrated response (depicted in red), respectively. The image ray connects a depth point with the surface position of its image. Image rays relate to migrated reflections in a similar way as a normal incidence rays relate to stacked primary reflections. Migrated primary reflection can be obtained by tracing rays vertically down from the free surface till the image time t_I is used.

$$\frac{\partial \mathbf{x}}{\partial \tau} = v^2 \mathbf{p},$$

$$\frac{\partial \mathbf{p}}{\partial \tau} = v \nabla \frac{1}{v},$$

(4.6)

with the initial conditions $\mathbf{x} = (x_1, x_2, x_3) = \mathbf{x}_0$ and $\mathbf{p} = 0$. The parameter τ is the parameter along the rays. In 3D media the system may be parametrized by two parameters α_1 and α_2 which represent ray coordinates. Furthermore, we consider the natural orthogonal basis defined as followed

$$\vec{e}_1 = \mathbf{p} \frac{1}{|\mathbf{p}|},$$

$$\frac{d\vec{e}_r}{d\tau} = -v \left(\frac{\partial \mathbf{p}}{\partial \tau} \cdot \vec{e}_r \right) \vec{e}_1 \ (r = 2, 3).$$

(4.7)

Because of the definition of the coordinate system, the derivatives $\partial \tau / \partial \alpha_{r,s}$, $(r, s = 2, 3)$ are equal to zero. Thus, we can establish the following relation:

$$\left(\mathbf{p}, \frac{\partial \mathbf{x}}{\partial \alpha_{r,s}} \right) = 0.$$

(4.8)

Now, we can introduce two matrices $\mathbf{Q}$ and $\mathbf{P}$ such that

$$\frac{\partial \mathbf{x}}{\partial \alpha_r} = \mathbf{Q}_{sr} \vec{e}_s \ (r, s = 2, 3),$$

$$\frac{\partial \mathbf{p}}{\partial \alpha_r} = \mathbf{P}_{sr} \vec{e}_s + \left(\frac{\partial \mathbf{p}}{\partial \alpha_r} \cdot \vec{e}_1 \right) \vec{e}_1 \ (r, s = 2, 3).$$

(4.9)

64

Let us look at the derivatives of $\mathbf{Q}$ and $\mathbf{P}$ with respect to the ray parameter τ:

$$
\frac{d\mathbf{Q}}{d\tau} = \frac{d}{d\tau}\left(\frac{\partial x_i}{\partial \alpha_r}(\vec{e}_s)_i\right) = \frac{d}{d\tau}\left(\frac{\partial x_i}{\partial \alpha_r}\right)(\vec{e}_s)_i +
$$
$$
+ \frac{\partial x_i}{\partial \alpha_r}\frac{d(\vec{e}_s)_i}{d\tau} = v^2\frac{\partial p_i}{\partial \alpha_r}(\vec{e}_s)_i = v^2\mathbf{P}
$$

$$(4.10)$$

and

$$
\frac{d\mathbf{P}}{d\tau} = \frac{d}{d\tau}\left(\frac{\partial p_i}{\partial \alpha_r}(\vec{e}_s)_i\right) = \frac{d}{d\tau}\left(\frac{\partial p_i}{\partial \alpha_r}\right)(\vec{e}_s)_i +
$$
$$
+ \frac{\partial p_i}{\partial \alpha_r}\frac{d(\vec{e}_s)_i}{d\tau} = -\frac{1}{v}\frac{\partial^2 v}{\partial x_m \partial x_i}\frac{\partial x_m}{\partial \alpha_r}(\vec{e}_s)_i =
$$
$$
-\frac{1}{v}\frac{\partial^2 v}{\partial x_m \partial x_i}Q_{pr}(\vec{e}_s)_i(\vec{e}_p)_m \equiv V_{sp}Q_{pr} = \mathbf{VQ},
$$

$$(4.11)$$

with the matrix $\mathbf{V}$

$$
\mathbf{V} = -\frac{1}{v}\frac{\partial^2 v}{\partial x_m \partial x_i}(\vec{e}_s)_i(\vec{e}_p)_m. \tag{4.12}
$$

Using matrices $\mathbf{Q}$ and $\mathbf{P}$, the 2×2 matrix $\mathbf{M}$ of second derivatives of traveltime with respect to spacial coordinates x_i can be determined.

$$
\mathbf{M} = \frac{\partial^2 \tau}{\partial x_m \partial x_i}(\vec{e}_r)_i(\vec{e}_s)_m = \mathbf{PQ}^{-1} \tag{4.13}
$$

Finally, we obtain the fundamental matrix for dynamic ray tracing system for the image rays which reads as

$$
\frac{d}{d\tau}\begin{pmatrix}\mathbf{P}\\\mathbf{Q}\end{pmatrix} = \begin{pmatrix}\mathbf{0} & v^2\mathbf{I}\\\mathbf{V} & \mathbf{0}\end{pmatrix}\begin{pmatrix}\mathbf{P}\\\mathbf{Q}\end{pmatrix} = \mathbf{S}\begin{pmatrix}\mathbf{P}\\\mathbf{Q}\end{pmatrix} \tag{4.14}
$$

This fundamental matrix is a matrix of special solutions for particular initial conditions. In our case, the slowness is known and the rays are in the origin. Therefore, we need normalized point-source initial condition. These initial conditions correspond

65

to solutions of the dynamic ray-tracing system for initial condition as follows

$$\begin{pmatrix} \mathbf{P}(0) \\ \mathbf{Q}(0) \end{pmatrix} = \begin{pmatrix} \mathbf{0} \\ \mathbf{I} \end{pmatrix} \Rightarrow \begin{pmatrix} \mathbf{P}_2(\tau) \\ \mathbf{Q}_2(\tau) \end{pmatrix} \tag{4.15}$$

The second derivatives of the traveltime field corresponding to a point source are given by

$$\mathbf{M} = \mathbf{P}_2\mathbf{Q}_2^{-1} \tag{4.16}$$

This allows us to approximate second-order traveltimes of a specified wave at arbitrary points near a reference ray by dynamic ray-tracing along that ray. If a point on the reference image ray is specified by $\mathbf{x} = x_i$, $i = 1,2$, the second-order traveltime approximation at point $\mathbf{x} + \Delta\mathbf{x}$ is available and given by

$$t(\mathbf{x} + \Delta\mathbf{x}) = t(\mathbf{x}) + \mathbf{p}\Delta\mathbf{x} + \frac{1}{2}\Delta\mathbf{x}^T\mathbf{M}\Delta\mathbf{x}, \tag{4.17}$$

where $\mathbf{M}$ $(M_{ij} = \partial^2 t/\partial x_i \partial x_j$, $i,j = 1,2)$. If we square this expression, keep only the terms up to second order in $\mathbf{x}$, and take into account the initial conditions for image rays, we finally obtain the traveltime approximation for paraxial image rays in the vicinity of the central image ray as read

$$t^2 = t(\mathbf{x})^2 + t(\mathbf{x})\,\Delta\mathbf{x}^T\mathbf{M}\Delta\mathbf{x}. \tag{4.18}$$

So far we assumed that the kinematic attributes of image rays are available. Now I will briefly discuss how the attributes can be extracted from the data.

To obtain image-ray wavefront curvatures, we propose to use CSP gathers (see chapter 2). In the single CSP gather, the following hyperbolic traveltime approximation is valid (see Equation 3.39):

$$t^M(h)^2 = (t_0^M)^2 + \mathbf{M}_{hh}h^2. \tag{4.19}$$

Comparing this equation with the traveltime approximation for paraxial image-rays given by Equation 4.18, we see that the matrix $\mathbf{M}_{hh}$ is linked to the wavefront curvature of the image-rays

$$\mathbf{M} = \mathbf{P}_2\mathbf{Q}_2^{-1} = \frac{1}{t_0}\mathbf{M}_{hh}, \qquad (4.20)$$

where we consider time t_0 to be the two-way traveltime along the central image ray.

In the next section, I will present a technique to construct smoothed velocity models based on the curvature of the image rays.

4.3 Methodology of Image Ray Tomography

In this section, we will consider the 2-D case. The propagation of the paraxial rays in the vicinity of the central image ray can be associated with a hypothetical wave due to a point source at the IIP. This wave is referred to as Image Incident Point (IIP) wave. The focusing of this wave at zero traveltime at the IIP is equivalent to focusing time-migrated reflection traveltimes at IIP's. Note that in contrast to the CMP traveltimes time-migrated reflection traveltimes are dip independent, thus, can be exactly described in terms of the IIP wave. A model is consistent with the data if all IIP waves focus at zero traveltime when propagated back into the subsurface. Figure 4.3b compares data vectors for NIP and image-ray tomography, respectively. In contrast to the NIP-wave tomography, the first-order traveltime derivatives are not required to construct the data vector since the phase slownesses of image rays are normal to the acquisition line. The image ray is solely determined by its wavefront curvature at the acquisition line.

The data vector required for the inversion consist of one-way zero-offset traveltimes T_0^M along the central image ray and the second spatial traveltime derivatives M_h at the respective image ray emergence location x_0

$$(T^M, M_h, x_0)_i, \; i = 1, ..., n. \qquad (4.21)$$

The data are extracted from the stacked CSP data at n-data pick locations by an

67

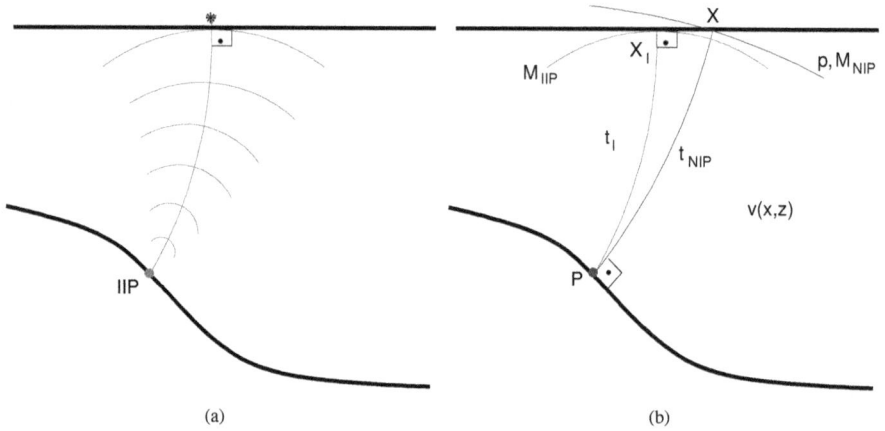

Figure 4.3: (a) The propagation of the paraxial rays in the vicinity of the central image ray can be associated with a hypothetical wave due to a point source at the IIP. This wave is referred to as Image Incident Point (IIP) wave. (b) Definition of data-vector components for 2-D NIP-wave and image-ray inversion. The data vector for NIP-wave tomography describes the second-order traveltime curve associated with an emerging wavefront of the normal ray and contains (T, M_{NIP}, p, x). The data vector for image ray tomography describes the second-order traveltime curve associated with an emerging wavefront of the image ray and consists of a data triple (t^M, M_{IIP}, x). The model components are the spatial location of the depth point P and the B-spline coefficients of the velocity field $v(x, z)$. The depth point is commonly referred to as the NIP point in the NIP-wave tomography or the IIP point in the image-ray tomography. X represents CMP location where the NIP-wave is measured. The X_I represents the CSP location where the IIP-wave is measured.

68

automatic picking procedure (Klüver and Mann, 2005). Each of these data points is associated with a corresponding IIP in the subsurface, characterized by its spatial location $(x,z)^{(IIP)}$.

The 2-D velocity model is described by two-dimensional B-splines

$$v(x,z) = \sum_{j=1}^{n_x} \sum_{k=1}^{n_z} m_{jk} \beta_j(x) \beta_k(z), \qquad (4.22)$$

where m_{jk} are B-spline coefficients representing the velocity model on a rectangular grid n_x and n_z are the chosen numbers of grid points in the horizontal and vertical directions. For the 2-D tomographic inversion, the model is therefore defined by the model parameters

$$(x,z)_i^{(IIP)}, \quad i = 1, ..., n_{data}$$
$$v_{jk} \quad j = 1, ..., n_x; k = 1, ..., n_z$$

$$(4.23)$$

The forward modeling of the quantities

$$(T^M, M_{IIP}, x)_i^{mod}, \quad i = 1, ..., n \qquad (4.24)$$

during the inversion process is performed by applying 2-D kinematic and dynamic ray tracing. Using kinematic ray tracing yields the emergence location ξ_0 of the image ray, while integration of equation 4.7 along the image ray yields the traveltime T_0^M.

It has been shown above that the parameters describing a second-order approximation of the traveltimes of emerging IIP wavefronts can be extracted from the CSP data. The kinematic wavefield attributes of time-migrated reflections describe the emerging hypothetical IIP wavefront in terms of second-order traveltime derivatives. In 2-D orthonormal coordinates, the second spatial derivative of the IIP-wave traveltime on the central image ray is given as,

$$M_{IIP} = P_2 Q_2^{-1}.$$

69

The inverse problem can be formulated as follows: a model vector $\mathbf{m}$, consisting of velocity model parameters m_{jk}, $j = 1,...,n_x$, $k = 1,...,n_z$ and image ray starting parameters at IIP, $(x,z)_i$, $i = 1,...,n$, is sought, that minimizes the misfit between a data vector $\mathbf{d}$, containing the picked values, and the corresponding modeled values $\mathbf{d}_{mod} = \mathbf{f}(\mathbf{m})$. The operator $\mathbf{f}$ symbolizes the dynamic ray tracing. The least-square norm is used as a measure of misfit (Tarantola, 1987). The modeling operator $\mathbf{f}$ is non-linear. Instead of solving a global nonlinear optimization problem the inverse problem is solved in an iterative way by locally linearising $\mathbf{f}$ and applying linear minimization during each iteration. The required Fréchet derivatives of $\mathbf{f}$ at the current model

$$\frac{\partial \left(T^M, M_h, x \right)}{\partial \left(x, z, v \right)}$$

are calculated during ray tracing by applying ray perturbation theory (Farra and Madariaga, 1987; Červený, 2001).

4.4 Synthetic example

As a first test, the image-ray based tomographic algorithm was applied to a synthetic example. For this purpose I used the same data as in the previous chapter (see Figure 3.6a which displays a simple synthetic model with an anticline in the middle.)

After the application of the CMRE stack, 592 data points (T^M, M^M, x) were picked in the resulting simulated ZO time-migrated section and the section containts wavefront curvature of the image rays. These served as input for the inversion. The velocity model consists of 104 B-spline knots: 8 knots in the x-direction with a spacing of 1000m and 13 knots in the z-direction with a spacing of 200m. The starting model was chosen to consist of a near-surface velocity of 1500 m/s and a vertical velocity gradient of $0.6\ s^{-1}$. To find the initial ray starting positions of the image ray in the subsurface, rays corresponding to locations of all data points were simply traced downward in the initial model normal to the acquisition line until the one-way traveltime T^M was reached.

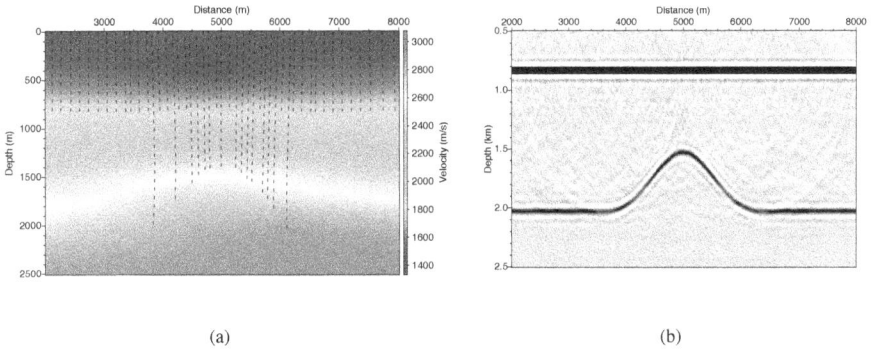

Figure 4.4: (a) Inversion result of the image-ray tomography. The reconstructed velocity model is displayed with some image rays. The input data for the inversion were picked from the CMRE stacked section and associated CMRE-attribute sections. (b) The depth-migrated section of CMP data using the estimated tomographic model.

The inversion result after 11 iteration is shown in the Figure 4.4a. The calculated image rays are also depicted. One can observe that the image rays are normal to the surface. Note that the model obtained as a result of the tomographic inversion is described by smooth spline functions, while the true model is blocky. The resolution of the obtained model depends on the grid spacing. However, the inversion result should be kinematically equivalent to the true model for all reflection events in the data. Figure 4.4a shows that the reconstructed model resembles a smoothed version of the true velocity distribution. The resulting smooth velocity model was finally used to perform prestack depth migration to investigate whether reflectors in the model are correctly imaged. For that purpose, a Kirchhoff migration algorithm based on eikonal traveltimes was used.

Figure 4.4b shows a stack of all common-offset migrations between 0 and 2000 m offset. Before the stacking, the Common Image Gathers (CIG) were muted according to Figure 3.6b. As expected from the results of Figure 4.5, all reflectors in the migrated image are correctly positioned. The bow-tie events in the seismic data due to the anticline structure were correctly unfolded.

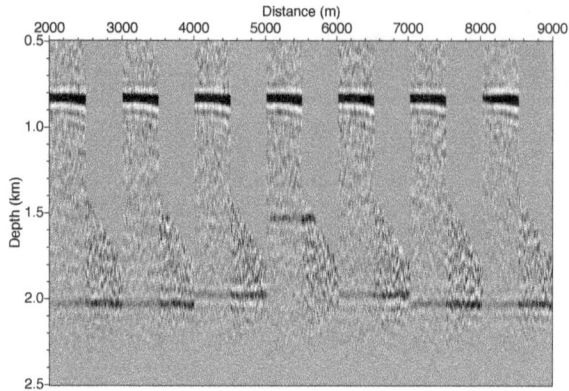

Figure 4.5: CIGs after prestack depth migration.The offsets displayed in the CIGs range from 0 to 2000 m. At shallow depths and large offsets, a mute has been applied to remove events with excessive wavelet stretch. The events in the CIGs are almost flat.

Figure 4.5 shows some CIGs at regularly spaced image locations with a separation of 1000 m. Each CIG represents the migration result at the respective image location as a function of offset. The offsets displayed in the CIGs in Figure 4.5 range from 0 to 2000 m. At shallow depths and large offsets, a mute has been applied to remove events with excessive wavelet stretch.

The velocity model is consistent with the seismic data if the results of prestack depth migration are kinematically independent of offset. The events in the CIGs displayed in Figure 4.5 are almost flat. This confirms that the model obtained with the proposed tomographic inversion is kinematically correct and suitable for further velocity model building techniques like migration velocity analysis.

The synthetic data example presented in this section shows that the concept of using the wavefront curvatures of the image rays for the construction of smooth, laterally inhomogeneous velocity model leads to a velocity model that is consistent with the data. The applicability of the presented method is limited since input data for the inversion are model-based. The inversion depends on the accuracy of the time-migration velocities used for the generation of CSP data. In case of strong lateral velocity variations, it is limited by the size of the offset aperture and by the strength of the velocity variation. Good result are achieved only if the hyperbolic assumption is applicable, i.e., for moderate lateral velocity variation in the seismic data.

Similar restrictions apply to the NIP-wave tomography. The major advantage of the image-ray tomography against the NIP-wave tomography is the unfolded wavefield in the input data and the focused diffractions which allow an easier and more reliable picking of the input data for inversion.

4.5 Concluding remarks

I have presented a new tomographic method for building smooth velocity models for depth imaging based on the wavefront curvature of the image rays. The image-ray propagation can be described in terms of a hypothetical IIP-wave. The IIP-wave is a wave which starts to propagate in the subsurface when a point source explodes at the IIP and arrive normally at the surface. Curvatures of the IIP waves can be directly extracted from common scatterpoint gathers by NMO like stacking procedure. The common scatterpoint gathers are prestack time-migrated data with a hyperbolic moveout which is based on the surface distance from a scatterpoint location to a collocated source and receiver. One of the advantages of CSP data is the absence of diffractions and triplications. Therefore, the estimation of attributes in the time-migrated domain might be more reliable than in the prestack CMP data.

The method can be seen as an additional tool to provide constraints for kinematic velocity model building. It is particularly useful in the areas where diffractions and triplications, which are located close to reflections, generate conflicting dip situation.

5 Imaging of seismic diffractions

In previous chapters, I have described several methods how to enhance reflection imaging based on Huygens hypothetical diffractions since conventional seismic processing is still tuned to image reflected waves. Recorded diffracted waves are lost by application of the proposed methods. However, most recently the seismic comunity has recognized the importance of the diffracted waves in seismic imaging. Only diffractions can provide reliable information on geological discontinuities like small-size objects, faults, rifts, zones of structural dislocations, fronts of metamorphism, or fracture corridors. They also can serve as quality control for velocity models in migration methods. Unfortunately, diffraction imaging is a big challenge because diffractions have much weaker amplitudes than reflections and often are masked behind the reflections. In order to improve the imaging of diffractions, several approaches have been developed over the last decade. The separation of the seismic data into diffraction-only data is the first step. These data can be used for velocity model building since a propper imaging of diffractions also strongly depends on the quality of the velocity model.

I propose an approach based on the Common-Reflection-Surface method to effectively separate reflection and diffraction events. The separation method is based on the stack of the coherent events using the Common-Reflection-Surface diffraction operator and applying a diffraction-filter during the generation of the stacked section. The diffraction-filter is based on the kinematic wavefield attributes which are obtained by the pragmatic search strategy. After separating seismic events I introduce a technique for migration velocity analysis using diffraction-only data based on the measure of the coherency along diffraction traveltimes in both time and depth domain. I also show an example illustrating the ability of the velocities extracted from diffractions to focus main reflections.

5.1 The main properties of diffracted waves

There are several approaches to define a real diffraction response since diffracted waves themselves differ in the nature. Commonly, a real diffraction response can be divided into two groups. The first group builds diffractions produced when the incident wavefield hits the edges, corners or vertexes of boundary surfaces. This responce is commonly referred to as diffracted waves. The second group builds diffractions produced when the incident wavefield passes objects or velocity perturbation about the characteristic size of the Fresnel zone, i.e., their size, d, is comparable to the apparent wavelength, λ, or with the Fresnel zone radius R_f, so the relations $d \sim \lambda$, or $d \sim R_f$ hold. Usually this response is referred to as scattered waves. Actually, depending on the seismic data, e.g., laboratory, reflection seismic, global seismology, the size d may vary from several millimeters to tens of kilometers. For the sake of simplicity, I will referr to both groups as diffracted waves. These structural objects mainly generating diffracted waves are illustrated in Figure 5.1. Figure 5.1a shows a single point-like diffractor in a homogeneous background. The single diffractor represents a point-like high-velocity inclusion. This geological object produces diffracted waves only if its dimension is comparable to the prevailing wavelength in the signal. Figure 5.1b shows an edge-like diffractor delineating a lateral sharp discontinuity in velocity. Figure 5.1c shows a single triangle structure simulating a cone-like diffractor. The process of seismic wave propagation in a general medium may be investigated with the help of the ordinary ray method (Červený, 2001). The theoretical consideration of the process of diffracted wave is a rather complicated problem. For its solution the diverse methods of theoretical and mathematical physics is utilized. The methods used in these problems can be treated from different points of view. An excellent overview of the methods can be found in (Gelchinsky, 1982). All methods can be divided into the following groups:

The first traditional group is formed by the analytically exact methods of solutions of

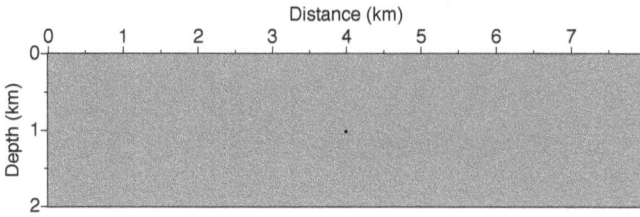

(a)

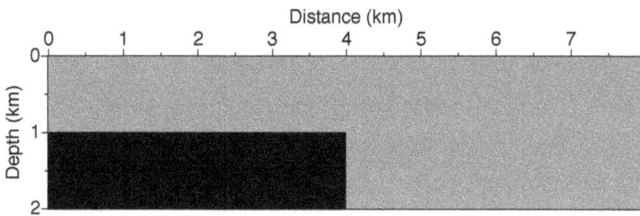

(b)

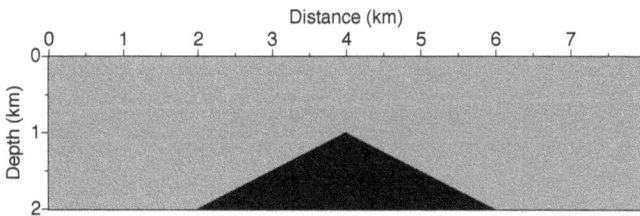

(c)

Figure 5.1: Three geological models mainly generate diffracted waves. (a) The single (point) diffractor in a homogeneous background. The single diffractor represents a high-velocity intrusion with small lateral and vertical extension. This geological object produces diffracted waves only if its dimensions are is comparable to the apparent wavelength in the signal. (b) The edge diffractor delineates a lateral sharp discontinuity in the velocity. (c) The single cone-like structure simulates cone-like diffractor.

76

the problems of scattering by the bodies of regular shape. The method of separation of variables lies in the basis of this group of methods. The obtained solutions are investigated analytically and numerically in order to receive the quantitative and qualitative consequences (Aki and Richards, 1980).

The second type of solutions can be called the integral equation method, because in the constructing of the solution the principal role is played by the integral equations appearing in it. Usually, the formulas of the Green-Kirchhoff type or of the field expansion into a complete or into a quasi-complete set of functions, taken from the problems with separable variables, lie in the basis of the methods of this group. To solve the integral equations it is suggested to use a set of various, often rather clever and elegant methods. The Wiener-Hopf technique, the matrix method and the method of optimal truncation are included in this group (Waterman, 1976). It is important to note that these are the numerical methods, because their physical and applied results are obtained after the realization of numerical iteration processes. These methods are especially convenient for low frequencies.

The solutions obtained with the help of perturbation theory are the third group of methods. The necessary condition of the constructive application of these methods is that the deviation of the scattering object parameters from the corresponding characteristics of the surrounding medium is small. The physical consequences are obtained from the qualitative and quantitative investigations of the constructed solutions (Aki and Richards, 1980).

The approximate local methods form the fourth group. Ray considerations and some locality principles lie in the basis of all modifications of these methods. The concept of a ray is generalized in these methods rather widely, (see, e.g., Keller, 1962; Berryhill, 1977; Kravtsov and Ning, 2010). The wavefield is presented as a asymptotic series constructed along the rays. In the range of validity the leading part of the field is determined by the first term of the series which have a clear physical meaning.

In general, assuming that the subsurface behaves as an acoustic medium, leads to the following summary concerning diffractions:

1. Diffraction amplitudes are frequency-dependent. High-frequency pulses excite less diffraction response than low-frequency pulses. Also, the waveform changes along the diffraction hyperbola because high-frequency components absorb first.

2. The maximum amplitude of an edge diffraction can attain one-half that of the associated reflection, and the diffracted waveform is identical to the reflected one. The maximum amplitude occurs on a seismic stacked section where the diffraction meets the reflection. This happens not at the apex of the diffraction hyperbola if the reflector is dipping.

3. The amplitude decrease is similar to the spherical divergence.

4. At the edge, the diffractions undergo a phase change of 180 degrees compared to reflections. This implies, the diffraction hyperbolas are divided into two regions in which the algebraic signs of the amplitudes are opposite. The part of the hyperbola off-end from the associated reflection has the same polarity as the reflection, while the part beneath the reflection has the opposite polarity.

5. Stacking velocities estimated to properly stack reflections, do not necessarily stack diffractions properly. This may be significant in the asymptotic region of a diffraction hyperbola, where the stacking velocities estimated for reflections are too high.

The above-mentioned conclusions are also visible from Figure 5.2.

In this section, I briefly reviewed mainly aspects of diffracted waves emphasizing the different behavior of them in amplitudes as well as kinematics in comparison to reflections. In the next section I will present a method how to describe the diffraction response and discuss several potential applications of diffraction in seismic processing.

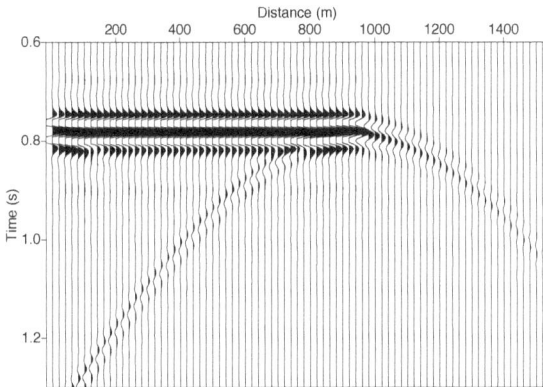

Figure 5.2: Seismograms generated for an edge-diffractor model. As predicted one can observe the phase reversal for the edge difffraction and also very low amplitudes compared to the reflection. Note that a gain was applied to emphasize diffraction amplitudes.

5.2 Extraction of diffracted events

As discussed in the introduction to the chapter, diffracted waves, allow to reliably image small-size subsurface features beyond the classical Rayleigh limit. Typical examples are small-size scattering objects, pinch-outs, fracture corridors, and karsts. Imaging of these features can be essential for the geological interpretation, e.g., in carbonate or salt environments (Krey, 1952; Kunz, 1960; Landa and Keydar, 1997; Kanasewich and Phadke, 1988).

Diffractions can also serve as quality control for velocity models in migration methods. A velocity model is assumed to be consistent with the data, if the diffractions are focused after migration. Moreover, instead using criterion of flatness of migrated common-image gathers commonly used in migration velocity analysis, it is possible to determine migration velocities with diffraction-focusing velocity analysis (Sava et al., 2005).

Diffraction imaging underwent a significant development in last decades. In the 70th and 80th of the last century the objective was the detection of diffracted waves in recorded data (Landa et al., 1987), or the determination of the diffractor location (Hubral, 1975). Nowadays, it is becoming an independent part of seismic processing and is aimed to correctly image the diffractions separately from the reflections (Moser and Howard, 2008). The separation is frequently based on the attenuation of the specular reflections in the recorded wavefield.

Fomel et al. (2006) separated diffractions in stacked sections using plane-wave destruction filters. The filters are prediction-error filters based on an implicit finite-difference scheme for the local plane-wave equation (Harlan et al., 1984). The criterion for separating diffracted and reflected events is the smoothness and continuity of local event slopes that correspond to reflection events. Khaidukov et al. (2004) proposed reflection-stack type of migration on prestack shot-gathers to focus reflections to a point and smearing diffractions over a large area. Muting the reflection focus and de-focusing the residual wavefield results in a shot gather that contains predominantly diffractions. Kozlov et al. (2004) proposed to image diffractions using the migration operator modified by an inclusion of a weight function which is constructed to suppress specular reflections. Their weight function is based on the anti-stationary phase filter. The migration using this filter uses only non-specular responses from the reflector outside the Fresnel zone. Reshef and Landa (2009) used the dip-angle domain to extract and analyze diffractions. The method is based on the construction of the dip-angle common-image gather. In this gather, after migration with the correct velocity, reflections appear as concave-shaped events while diffractions are flat. The different shape of the seismic events allows to separate them using a hybrid radon transform (Klokov et al., 2010). However, these methods are based only on the suppression of the considered reflections, and the assumption that residual seismic events are diffractions.

Berkovitch et al. (2009) proposed an other method for diffraction imaging. The method is based on an optimal summation of the diffracted events. Their method utilizes the diffraction multifocusing stack which uses a local-time-correction formula with two stacking parameters: the emergence angle and the radius of the curvature of the diffracted wavefront. The diffraction multifocusing stack separates diffracted and reflected energy in the stacked section by focusing diffractions to the diffraction locations and smearing the reflection energy over a large area.

I introduce an approach for diffraction imaging which combines the reflection attenuation and the coherent summation of diffracted events. The method uses the CRS technique (see chapter 3). According to the CRS theory, a diffractor is associated with a reflector segment with an infinite curvature and an undefined orientation (Mann, 2002). A reflector segment with infinite curvature implies that $R_N = R_{NIP}$ or $M_N = M_{NIP}$, respectively. This principle is illustarted in Figure 5.3. From the figure is it obvious that with decreasing reflector curvature the radii R_N and R_{NIP} become closer. As opposed to reflections, for diffractions any direction describes a possible zero-offset ray along which the NIP-wave and N-wave can be considered. For a diffractor every emerging ray is a 'normal' ray (Figure 5.3)

If the kinematic wavefield attributes for an arbitrary point in the data are known, an approximation of the prestack diffraction response is available by simply substituting R_{NIP} for R_N or vise versa in the conventional 2D CRS operator given by equation 3.24, i.e.,

$$t^2 = \left[t_0 + \frac{2 \sin \beta_0 m}{v_0} \right]^2 + \frac{2 t_0 \cos^2 \beta_0}{v_0 R_{NIP}} \left[m^2 + h^2 \right]. \tag{5.1}$$

Equation 5.1 represents the CRS-based diffraction (CRSD) operator which approximates the diffraction response up to second order. The CRSD traveltime surface is a hyperboloid in (t, m, h) space which is obtained by rotating a hyperbola around its semi-minor axis. The CRSD operator assigns the stacked result to the stationary point of the traveltime surface with respect to (m, h), which coincides with the ZO time t_0 of the 'normal' ray. The DSR operator assigns the stacked result to the operator apex which coincides with the ZO time of the image ray. The CRSD operator does not focus the diffractions to their apex but represents a fit to the traveltime based on a coherence criterion.

The CRSD operator is a single-square root operator thus an approximation valid for

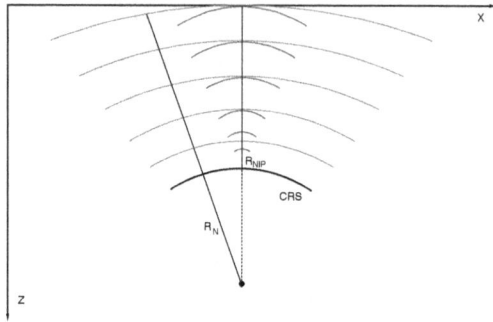

(a)

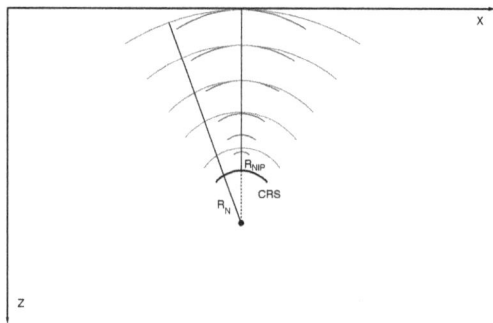

(b)

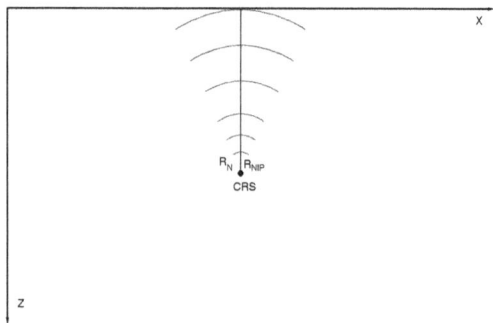

(c)

Figure 5.3: Illustration of the CRS approach for diffraced events. Red lines delineate normal wave, blue lines delineate NIP-wave. From the figure is it obvious that with decreasing reflector curvature the radii R_N and R_{NIP} become closer. As opposed to reflections, for diffractions any direction describes a possible zero-offset ray along which the NIP-wave and N-wave can be considered. For a diffractor every emerging ray is a 'normal' ray.

82

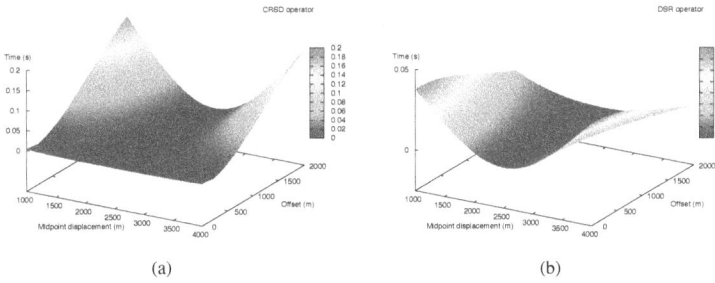

Figure 5.4: Traveltime-error plots for a 2D homogeneous model containing a single diffractor with an extension of 100 m at a depth of 1 km for the CRSD (a) and DSR operator (b), respectively (see Figure 5.5a). The velocity is 1500 m/s. The CRSD traveltime errors are negligible for short offsets and the plane defined by $m = 0$. For near offsets and large midpoint aperture the CRSD operator fits the diffraction response better than the DSR operator because the point-diffractor condition is not fulfilled here.

short offsets even in homogeneous media. Figure 5.4 shows traveltime-error plots for a 2D homogeneous model with a single diffractor at a depth of 1 km for the CRSD and DSR operator, respectively. The velocity is 1500 m/s. As I can observe in Figure 5.4a, the CRSD traveltime errors are negligible for the short offsets and the plane defined by $m = 0$. The larger the offset, the more the traveltime error increases. The traveltime-error plot for DSR operator is shown in Figure 5.4b. We also observe, for the near offsets and large midpoint aperture the CRSD operator fits the diffraction response better than DSR operator. For heterogeneous media, both operators are valid in the hyperbolic assumption. It cannot be quantified which operator may fit the data better because it is model/data dependent. To keep the offset small will improve the fit for both operators though.

It was mentioned above that the radii of curvature of the NIP-wave, R_{NIP}, and normal wave, R_N, coincide for diffractions. Thus, the ratio of R_{NIP} and R_N can be used to identify diffracted events (Mann, 2002). In the ideal case, a diffractor should yield a ratio of R_{NIP}/R_N equal to one. Strictly speaking, this applies within the high frequency limit since the operators are kinematic. Seismic data are always

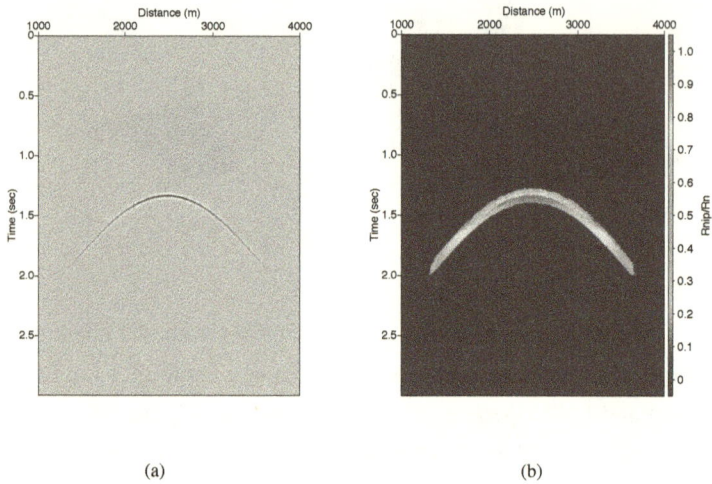

Figure 5.5: 2D homogeneous model with a small lens acting like diffractor at a depth of 1 km. The velocity is 1500 m/s. (a) The CRS stacked section of the model. (b) The ratio of radii R_{NIP}/R_N. I observe a decrease of the ratio with rising distance from the diffraction apex.

band limited and do not allow such simple distinction. Moreover, R_{NIP} and R_N are determined from data using an approximation to the real traveltime and it is not possible to determine diffractions in a binary way, i.e., $R_N = R_{NIP}$ is a diffraction and $R_N \neq R_{NIP}$ is a reflection. The ratio R_{NIP}/R_N for a simple homogeneous model is illustrated in Figure 5.5b.

The model contains of a small lens with lateral extension of 150 m simulating diffraction response. The velocity is 1500 m/s. I observe an increase of the ratio with rising distance from the diffraction apex. Since the radii of curvature slightly differ, a soft transition is required which can be achieved by thresholding. This transition should display a smooth and fast decay in order to sufficiently separate diffractions from reflections. I suggest the following function which serves as a guide for thresholding:

$$T_F(m_0, t_0) = e^{-\frac{|R_N - R_{NIP}|}{|R_N + R_{NIP}|}}. \tag{5.2}$$

This function is about one for R_{NIP} close to R_N and rather small if R_{NIP} and R_N differ. I weight the stacked result with one if the function $T_F(m_0,t_0)$ is above the threshold and with zero in the opposite case. The choice of the threshold depends on the complexity of the subsurface and the spectral content of the data. The lower the threshold, the more residual reflections will remain in the data. The application of the designed filter to poststack sections will not include reflected events in the stack because they have a lower value of $T_F(m_0,t_0)$. Diffracted events will remain in the stack because they have a higher value of $T_F(m_0,t_0)$. The resulting stacked section will then contain predominantly diffraction energy.

Subsurface structures with small radii of curvature with respect to the prevailing wavelength in the signal may appear very similar to diffraction events. Events from these structures will pass the filtering process and interfere with diffractions. In conflicting dip situations, reflected and diffracted events contribute to the same ZO location while they have different kinematic wavefield attributes. The kinematic wavefield attributes of both events should be considered to properly separate the seismic events. I propose to use an extended CRS stack strategy as described by Mann (2002) to avoid this potential problem. The strategy allows us to detect conflicting dip situations estimating kinematic wavefield attributes separately. All attributes are then used for the filtering process.

Depending on the value of the threshold, residual reflections may be present in the resulting stack. These reflections may still have greater amplitudes than diffractions and the diffraction amplitudes may be distorted by the filtering process in conflicting dip situations. To enhance the diffraction amplitudes and suppress the residual reflections, I apply the CRSD operator to the poststack section utilizing the partial CRS stack (Baykulov and Gajewski, 2009). For a sample with a certain CMP and time coordinate, I calculate the diffraction traveltime curve using the poststack CRSD stack operator given by

$$t^2 = \left[t_0 + \frac{2\sin\beta_0}{v_0}m \right]^2 + \frac{2t_0\cos^2\beta_0}{v_0 R_{NIP}}m^2, \tag{5.3}$$

85

and stack the amplitudes within a large midpoint aperture. For a point belonging to the residual reflected event, the poststack CRSD operator will stack the amplitudes coherently only along the segment of the residual reflection. Thus, stacking along a large diffraction trajectory will sum up both coherent and incoherent events. The residual reflected events will be further suppressed because of the destructive interference. For a point belonging to the diffracted event, the CRSD operator stacks the amplitudes coherently along the whole diffraction trajectory. The diffracted event will be enhanced because of constructive interference (see Figure 5.7).

5.3 Velocity model building using diffractions

After separating diffractions from reflections, I use the diffraction-only data for poststack velocity analysis. The conventional stacking velocity analysis is tuned to reflections which have another amplitude decay and waveform as diffractions. For the sake of simplicity, let us assume edge diffractions. The diffracted waveform coincides with the reflected waveform only exact above the edge and the diffraction amplitude is half of the reflection one. The amplitude decay cannot be explained by geometrical spreading. Therefore, the offset aperture within we usually apply the stacking velocity analysis is not suitable for diffractions. In the separated section, we have predominantly diffracted energy, therefore, the velocity analysis can be tuned to diffractions.

How can we use the separated diffractions for velocity model building? We know, e.g., that poststack time migration focuses diffractions to their apex if the velocity model is correct. Poststack velocity model building for diffractions based on the focusing criterion is described, e.g., in Harlan et al. (1984) and Fomel et al. (2006). Their methods incorporate repeated migrations of the data using different migration velocities and subsequent determination of the best velocity by a focusing analysis.

I introduce another technique to build a time-migration velocity model based on a coherence analysis for diffraction curves. As a measure of the coherency I use the semblance norm but others are also possible.

Poststack time migration velocity analysis

The diffraction traveltime curves are determined by the DSR operator given by Equation 2.18. For the poststack section, the half-offset is equal zero, i.e., $h = 0$. Thus, the DSR operator simplifies to the hyperbola

$$t_D = \sqrt{t_0^2 + \frac{4m^2}{v^2}} \tag{5.4}$$

where v is the time-migration velocity, m is the midpoint displacement with respect to the considered CMP position, t_0 corresponds to the ZO two-way traveltime. The DSR operator fits the diffracted event in case the time-migration velocity is correct. The coherence analysis will provide a high semblance value in this case. For each sample in the stacked section, I perform a velocity scan from low to high velocities, i.e., for each velocity I compute the diffraction curve using the poststack DSR operator given by Equation 5.4, stack the amplitudes along this traveltime curve, and determine the semblance value. The output is a migration-velocity panel that is suitable for picking time-migration velocities. The velocity with the highest semblance value will be considered as the sought migration velocity.

Since the DSR operator is defined at its apex time, the coherence analysis will provide the highest semblance value for the correct migration velocity and for the apex location of the diffraction traveltime. This additional information can be used in the stacking procedure. The CRS approach assumes the continuous surface around the NIP. That means that for the NIP located in the nearest vicinity of the fault, edge, or truncations, the CRS approach may results in a smeared image. The stacking aperture for a Common-Reflector-Surface element should either start or terminate at the diffracting element, i.e., faults, edge, or pinch out. In the velocity analysis of the diffraction-only data, I identify the apex position of the diffracting subsurface feature that allows us to apply optimized CRS apertures which will exclude these regions.

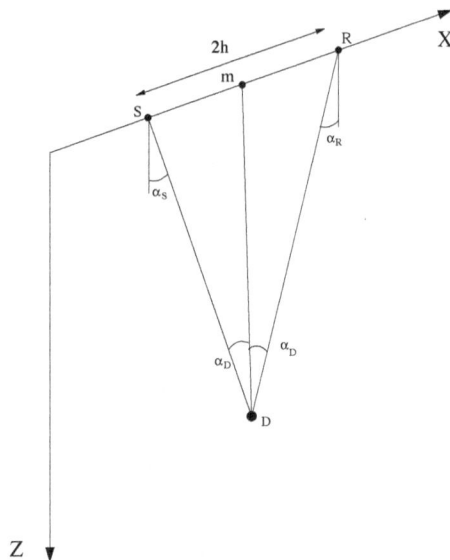

Figure 5.6: 2D cartoon illustrates traveltimes along the ray segments for a diffractor embedded in a medium with constant velocity.

Poststack depth migration velocity analysis

For the determination of a depth-velocity model, I propose a velocity scan based on the semblance analysis for the diffraction traveltimes. The traveltimes are usually calculated by ray tracing. However, for simpler types of media, there is an analytical expression given by the Kirchhoff approach. For some specific velocity distributions, e.g., constant velocity, constant velocity gradient, fully analytical solution of the Kirchhoff integral can be derived. If the velocity distribution in the true earth model can be reasonably well represented by one of the above-considered situations, a fast velocity analysis can be performed.

The traveltime from one depth point under consideration to a receiver or source in an inhomogeneous medium, i.e., the velocity is a function of depth only, $v = v(z)$, can be computed by means of the sum of the traveltimes along the ray segments SD and DR (see Figure 5.6). For the sake of simplicity the 2D case is illustrated in the Figure 5.6.

88

In the general 3D case, the traveltime along the in-plane ray SDR is the integral (Martins et al., 1997)

$$t_D(\xi;x,z) = \frac{1}{v_0} \left(\int_0^z \frac{n^2(z')dz'}{\sqrt{n^2(z') - \sin^2 \alpha_S}} + \int_0^z \frac{n^2(z')dz'}{\sqrt{n^2(z') - \sin^2 \alpha_R}} \right) \qquad (5.5)$$

The unknown starting angles α_S and α_R in the above integral can be eliminated from the integration result by inverting the following integral for the horizontal distance between point S or R and D,

$$x - x_i = \sin \alpha_D \int_0^z \frac{dz'}{\sqrt{n^2(z') - \sin^2 \alpha_i}} \quad (i = S, R)$$

Using Snell's law for diffracted rays (see e.g., (Keller, 1962)) as well as combining both equations above it is possible to establish the following formula

$$\cos \alpha_D = \sqrt{1 - v(D)^2 \left(\frac{x - x_i}{\sigma_i}\right)^2} \quad (i = S, R) ,$$

where σ_i is a parameter describing out-of-plane effects. It reads

$$\sigma_i = V_0 \int_0^z \frac{dz'}{\sqrt{n^2(z') - \sin^2 \alpha_i}} \quad (i = S, R) \qquad (5.6)$$

Now we can consider two cases: the case of a constant velocity and the case of constant velocity gradient.

Constant velocity

The ray segment SD and DR are straight lines. Denoting their respective length by l_S and l_G, i.e.,

$$l_i = \sqrt{(x - x_i)^2 + z^2} \, (i = S, R) \qquad (5.7)$$

89

integral 5.6 is solved yielding the result $\sigma_i = v_0 l_i \, (i = S, R)$ The traveltime is the well known DSR equation, i.e.,

$$t(x,z) = \frac{1}{v_0} \left(\sqrt{(x-x_S)^2 + z^2} + \sqrt{(x-x_R)^2 + z^2} \right)$$

or in midpoint coordinates

$$t(x,z) = \frac{1}{v_0} \left(\sqrt{(x-h)^2 + z^2} + \sqrt{(x+h)^2 + z^2} \right) \tag{5.8}$$

Constant vertical gradient of velocity

Now the velocity distribution is controlled by the following law:

$$v(z) = v + \gamma z, \tag{5.9}$$

where γ can be considered as an inhomogeneity factor. Using this velocity distribution in integrals 5.5 and 5.6, it is possible to express the parameters σ_S and σ_R as

$$\sigma_i = \frac{l_i}{2} \sqrt{\gamma^2 l_i^2 + 4 v v_0} \, (i = S, R),$$

where v is the velocity at D and l_S and l_G being the distances from S and R to D as defined in equation 5.7. Once the depth point is specified, all quantities are known to calculate the traveltime. Substituting the velocity distribution 5.9 into integral 5.5 yields (Martins et al., 1997)

$$t(x,z) = \frac{1}{\gamma} \ln\left(B_S B_R\right), \tag{5.10}$$

where

$$B_i = 1 + \frac{\gamma^2 l_i^2 + 2\gamma \sigma_i}{2 v v_0} \, (i = S, R),$$

Similarly to the time-domain, we perform a gradient scan from low to high gradients. This means, for every depth point we compute diffraction traveltimes using Equation 5.10 evaluating the semblance value for different gradient values. The output is a

90

gradient panel suitable for picking. The picked gradients can be used to calculate the velocity distribution according to Equation 5.10. The calculated velocities represent smoothed interval velocities which are suitable as initial velocities for the tomographic inversion or depth migration.

5.4 Synthetic example

Figure 5.7a displays a stacked section of a synthetic model containing five layers and four small lenses which simulate diffractors. The velocity within the layers is constant. The velocity in the first layer is 1500 m/s, in the second layer 1580 m/s, in the third layer 1690 m/s, in the fourth layer 1825 m/s and in the fifth layer 2000m/s. Four small lenses with a lateral extension of 200 meters in the fourth layer produce diffractions. I used Seismic Un*x to generate synthetic seismograms with the Gaussian beam method. I used a Ricker-wavelet with a prevailing frequency of 25 Hz.

The following processing steps are performed: (1) the pragmatic search strategy to estimate the CRS attributes (Mann, 2002); (2) stacking of the prestack data using the CRSD operator to emphasize diffractions; (3) application of the diffraction filter to the stacked section to further suppress reflected events; (4) partial CRS stack with the poststack CRSD operator to enhance diffraction amplitudes and attenuate residual reflected events. After these steps, we obtain diffraction-only data, i.e., a stacked section that contains predominantly diffraction energy (Figure 5.7b). We used 0.9 as threshold for the filter. This threshold corresponds to $R_{NIP} \approx 0.8R_N$. As can be observed in Figure 5.7b, the diffracted events are well separated from the reflected events, even in the areas of conflicting dips.

The diffraction-only data we use for poststack time-migration velocity analysis. We applied coherence analysis based on the semblance norm for each sample in the poststack section. Figure 5.8a illustrates a migration-velocity panel for a CMP location directly above one of the diffractors. Figure 5.8b shows the coherence values

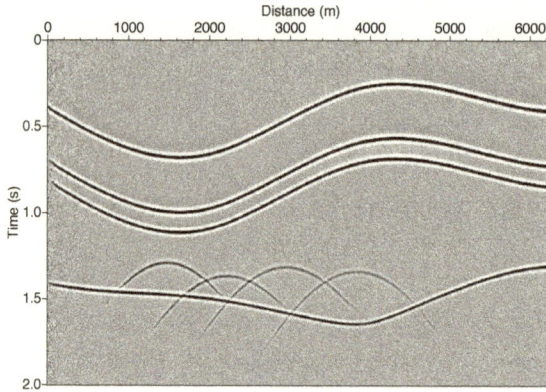

(a)

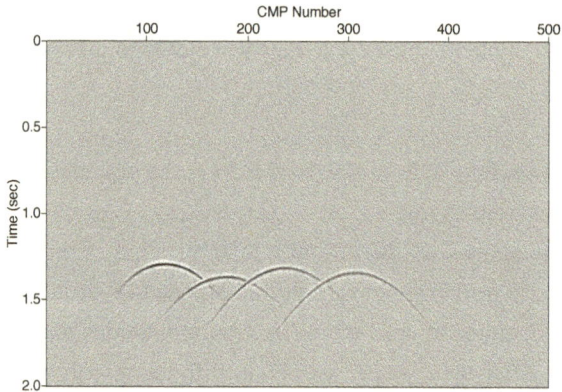

(b)

Figure 5.7: Synthetic example with four small lenses of 200 m lateral extension simulating diffractions. Stacked section of the whole recorded wavefield (a) and diffraction-only data (b). Lateral extension of the whole seismic line is 6250 m. Good separation of seismic events has been achieved. Conflicting dips are preserved.

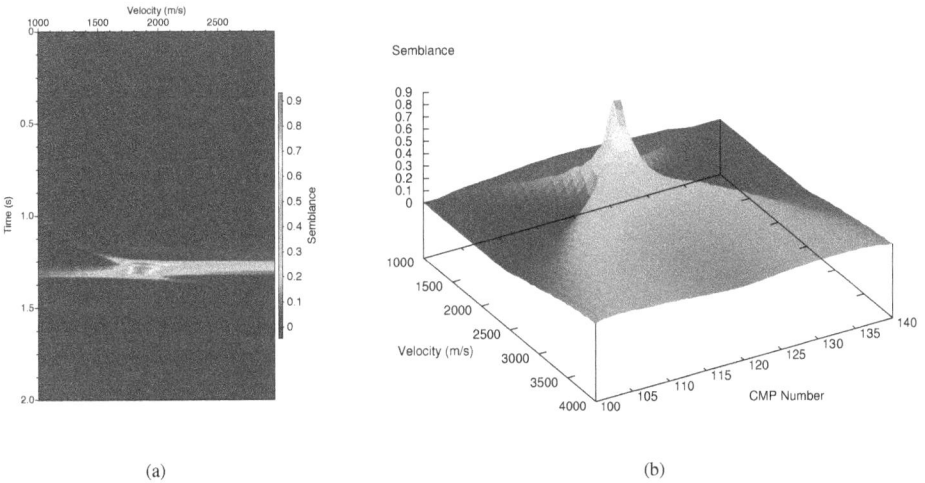

Figure 5.8: Results for the synthetic model with four small lenses. (a) velocity spectrum after poststack time-migration velocity analysis of a CMP located directly above one of the diffractors. Red color indicates high semblance. The corresponding distribution of the coherence as function of velocity for the same CMP is illustrated in (b). We observe a sharp and narrow maximum for the diffraction apex allowing easy picking of velocities.

as a function of velocity and CMP position. I observe a sharp and narrow maximum for the apex of the diffraction. The time was manually picked and corresponds to the maximum of the coherence value. Figure 5.11a shows the time-migration velocity model obtained by spline interpolation between picked locations.

Also, the diffraction-only data, Figure 5.7, are used for the depth-migration velocity analysis. Figure 5.9 shows a gradient panel for a CMP which is located directly above one of the diffraction apexes. Figure 5.11b shows the depth-migration velocity model obtained by spline interpolation between calculated velocities. The velocities are previously computed for the whole trace where the gradient has the maximum value of semblance.

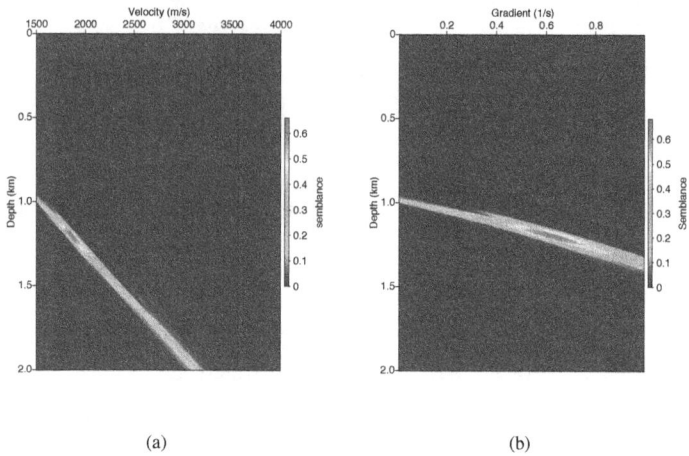

(a)   (b)

Figure 5.9: Results for the synthetic model with four small lenses. (a) velocity spectrum after poststack depth-migration velocity analysis using an average velocity. (b) velocity spectrum after poststack depth-migration velocity analysis using a gradient scan. Red color indicates high semblance.

Figure 5.10 shows the time-migrated sections obtained by Kirchhoff poststack time-migration using the RMS velocities (a) and the time migration velocities estimated from the diffraction-only data (b). The RMS velocities are obtained using conventional velocity analysis for the recorded wavefield. The diffractions are focused at the lenses in both cases. However, in the section obtained with the velocity model estimated from the diffraction-only data the diffraction energy is better focused, indicating a better velocity model.

Applicability of velocities extracted from diffractions to focus reflections

Although the velocities estimated using diffraction-only data are used to focus diffractions, they also can be used to focus main reflections. This model was used for Kirchhoff prestack time migration of the whole recorded data. Several Common-Image-Gathers corresponding to the apex positions of diffractions are shown in Figure 5.12a. The diffracted events are almost perfectly flat. Reflections from layers that are above the layer containing diffractors exhibit strong residual moveouts. However, the reflection from the layer containing diffractions displays horizontal moveout, particularly for near offsets. The final prestack time-migrated section is shown in Figure 5.12b. As expected from the CIGs, the first three reflections are distorted. The focusing of the last reflection and diffractions is better. This velocity model is used for Kirchhoff prestack depth migration of the whole recorded data. Several Common-Image-Gathers corresponding to the apex positions of diffractions are dispayed in Figure 5.13. The diffracted events are almost perfectly flat. With respect to reflections, I observe the same trend as in the time domain. Reflections from layers that are above the layer containing diffractors exhibit strong residual moveouts. The reflected event from the layer containing diffractions is almost flat, particularly for near offsets.

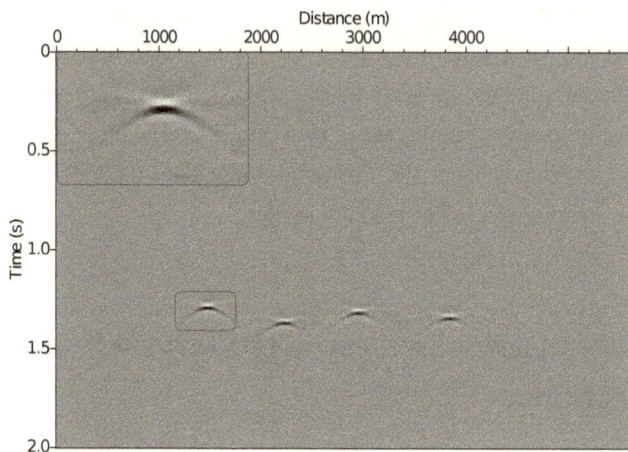

(a)

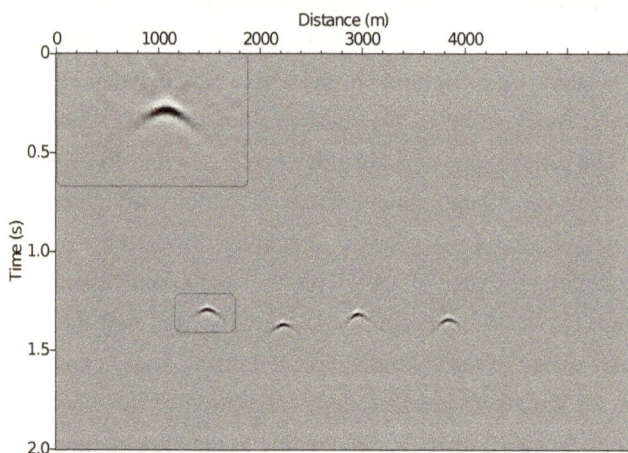

(b)

Figure 5.10: Imaging of the synthetic example with four small lenses. Lateral extension of the seismic line is 6250 m. Poststack time-migration of the diffraction-only data. Time-migrated sections: (a) obtained with the estimated RMS velocities and (b) obtained with the velocities estimated from the diffraction-only data. In the latter section the diffractions are better focused and their lateral extension is about 200 m as in the model.

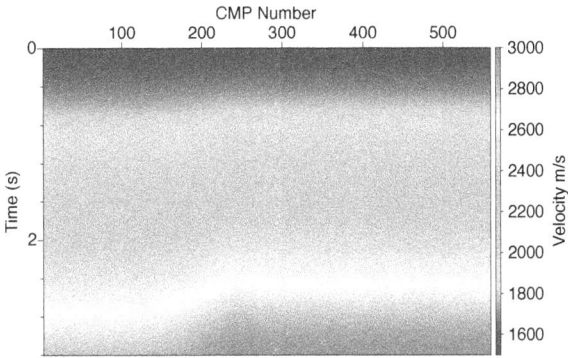

(a)

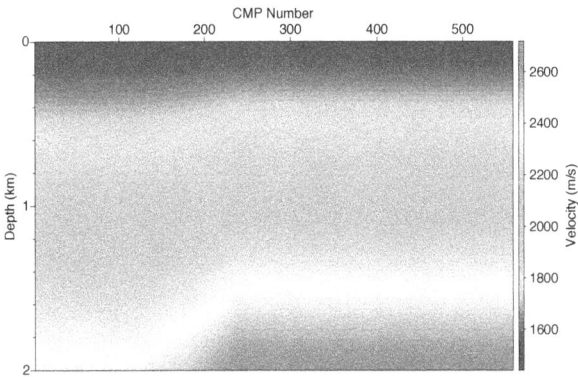

(b)

Figure 5.11: Results for the synthetic model with four small lenses. (a) Time-migration velocity model obtained by spline interpolation between picked locations. (b) Depth migration velocity model is obtained by spline interpolation between calculated velocities.

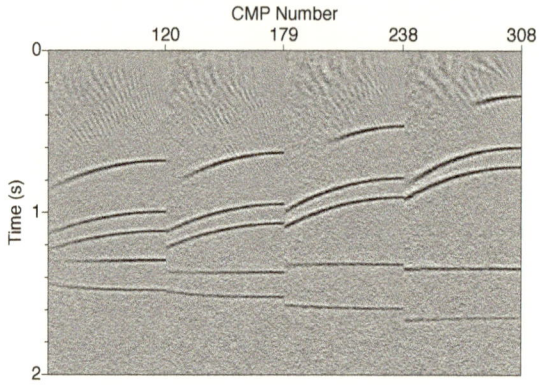

(a)

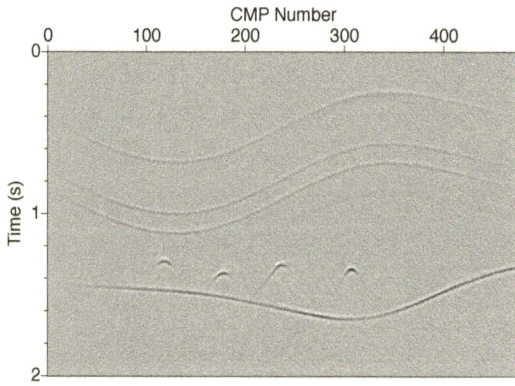

(b)

Figure 5.12: Imaging of the synthetic example with four small lenses. Common-Image-Gathers after prestack time migration for locations corresponding to the apex positions of diffractions. Time migration of the data was performed with velocities estimated from the diffractions.

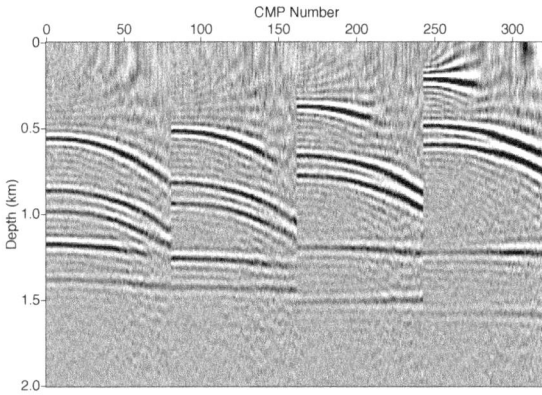

(a)

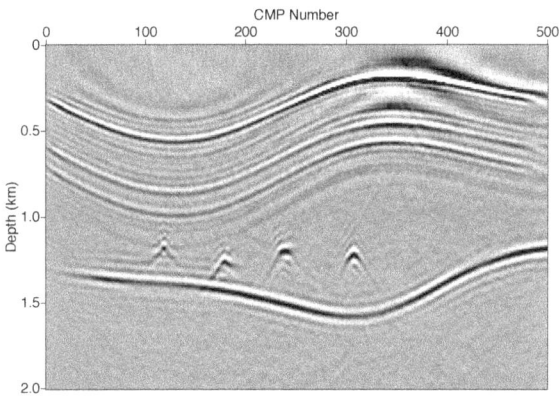

(b)

Figure 5.13: Imaging of the synthetic example with four small lenses. Common-Image-Gathers after prestack depth migration for locations corresponding to the apex positions of diffractions. Depth migration of the data was conducted with velocities estimated from the diffractions.

5.5 Concluding remarks

I have proposed a new method to image diffracted events. The method is based on the separation reflections and diffractions using CRS attributes. The process consists of stacking of the coherent events with a CRS-based diffraction operator followed by attenuation of reflected events in the poststack domain with a new type of filter. This filter is based on CRS attributes which are determined by the CRS approach. No additional analysis or search is required. In a post processing step, the diffraction amplitudes are enhanced using partial CRSD stacks. A subsequent time migration velocity analysis on the diffraction-only data provides time migration velocities which are then used for Kirchhoff poststack time migration.

The radii of the normal and NIP-wave allow to identify seismic events. For diffractions, both radii should coincide, i.e., their ratio is close to one. I used an exponential threshold function instead of the simple ratio to stabilize the filter process and to allow for a smooth transition from diffractions to reflections. The choice of the threshold controls the separation of seismic events and depends on the complexity of the subsurface as well as on the frequency content of the data. The lower the threshold, the more residual reflections will remain in the diffraction-only data. Because of possible distortion of the diffraction amplitudes in conflicting dip areas during the filtering, I apply a partial CRS stack for an enhancement of the diffractions.

Applications of the method to synthetic data confirm a good separation of seismic events followed by the presented method. The developed filter is not limited to stacking trajectories. The application to prestack data is also possible. Moreover, the extension of the method to 3D is straightforward (Dell and Gajewski, 2012).

The diffraction-only data can be used for a poststack migration velocity analysis. In the poststack domain, the data are reduced and a good S/N ratio is present. This allows fast time- and depth migration velocity analysis. The time-migration velocity analysis consists of a velocity scan for each sample in the stacked section and the evaluation of the semblance norm. The depth-migration velocity analysis utilizes a gradient scan for each sample and also the evaluation of the corresponding semblance norm. Applications of the method to synthetic data demonstrate that the determined velocities provide images with well focused diffracted events.

The migration velocities extracted from the diffracted events can also be used to image reflected events. The synthetic data application shows good focusing of the reflections from the layer containing diffractions. Focusing reflections is not sufficient if no diffractions are present in the corresponding layer. However, the field data usually contain a great deal of diffractions randomly distributed along the seismic profile. The velocity information extracted by the presented method from these diffractions can surely be used as an initial velocity model for further velocity analysis using the prestack data-volume.

6 Application to field data

In this chapter, I demonstrate the application of the developed imaging techniques to a marine field data example from the Eastern Mediterranean. The dataset was kindly provided by TGS. The data were chosen to obtain velocity information contributing to an elaboration of the respective geologic setting.

6.1 Study area, acquisition geometry and preprocessing

The data were collected in the central Levante Basin, located in the Eastern Mediterranean (see Figure 6.1), which is bounded to the South by the Egyptian and to the East by the Levante coast and reaches to the Cyprus Arc and the Eratosthenes Seamount in the North and West, respectively. The data cover the so-called Messinian Evaporites, i.e., relatively young salt formations. These Evaporites precipitated during the Messinian Salinity Crisis when the interaction of plate tectonics and eustatic sea level fall led to the closure of the gateway, strait of Gibraltar, between the Mediterranean and the Atlantic. In the Levante Basin, they reach a maximum thickness of about 2 km.(Netzeband et al., 2006b).

Recent publications show a complex seismic stratigraphy of the evaporite sequence (see e.,g., (Netzeband et al., 2006b; Cartwright and Jackson, 2008)), and several units can be traced throughout the entire basin. The deformation pattern of the intra-evaporite sequences including folds and thrust faulting gives evidence for extensive

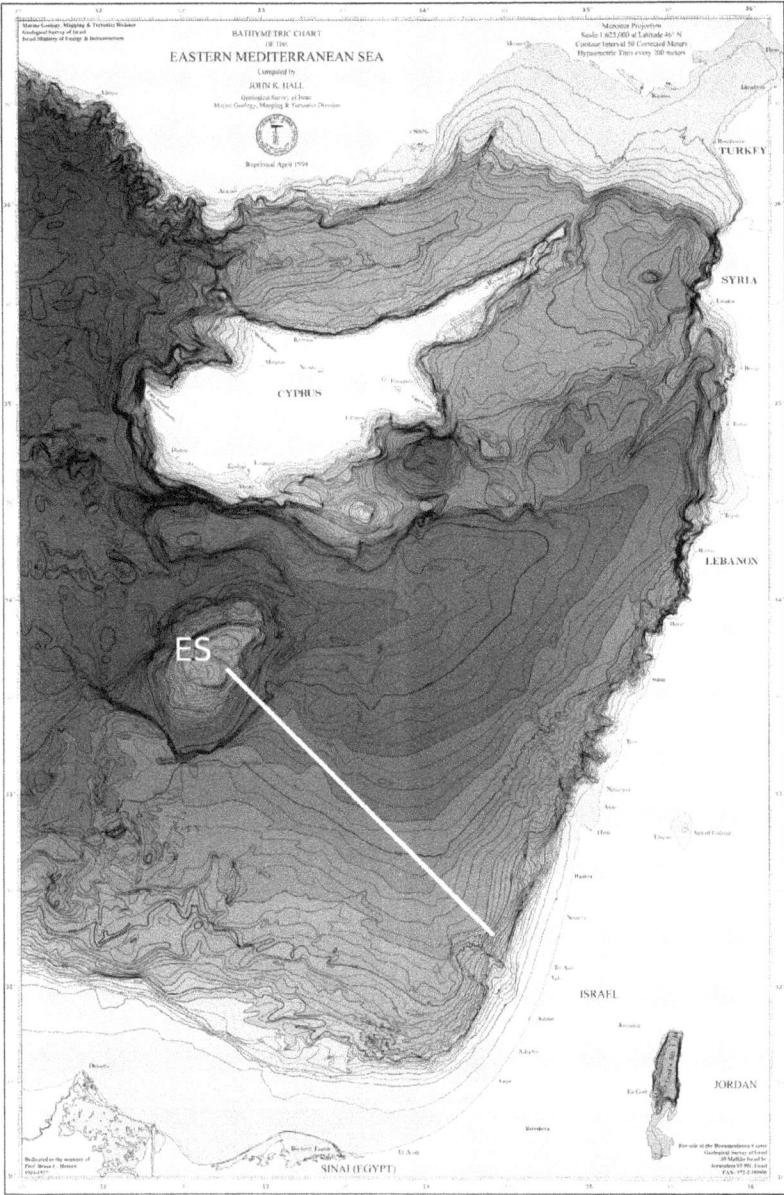

Figure 6.1: Map of the Levante Basin. The white line shows the location of the whole cross-section. A part of the section covering 2000 CMP locations was processed (see Figure 6.2). The Eratosthenes Seamount is denoted by ES.

103

salt tectonics during the depositional phase. However, two different mechanisms can be also considered, i.e., lateral creep of the evaporites which means basically gravity gliding, and plate tectonics.

It is commonly accepted that the load of the Nile Deep Sea Fan pushes the entire evaporite layer in a NNE direction parallel to the bathymetric gradient. In front of the lower fan as well as beneath the sediment prism off the Levante margin, the top of the evaporites declines towards the underlying conformity. This rollback results not only from the lateral creep of the salt but also from differential subsidence caused by the laterally varying thickness of the overburden (Netzeband et al., 2006a). Backstripping analysis showed that off the Southern Levante coast the (apparent) rollback structure vanishes after the sediment unload. However, the presence of salt rollers indicates that some extensional tectonics occurred (Gradmann et al., 2005; Hübscher and Netzeband, 2007). However the Messinian Evaporites may also be overprinted by plate tectonics (Neev, 1977; Neev et al., 1976; Neev, 1975). The recent activity and their impact on the structural evolution of the Levante margin is still a matter of debate. Usually, time-migrated seismic sections are used for structural interpretation. Owing to the high interval velocity of the Messinian layer, lateral thickness variations may pretend deep-rooted folds or faults. From time-migrated sections it is almost impossible to infer whether the lower boundary of the evaporites is an undisturbed detachment surface or whether it is dissected and offset by faults (Netzeband et al., 2006a).

In order to better understand the deformation pattern present in the Levante Basin and to elucidate the origin of the identified evaporitic facies of the intra-evaporitic sequences an optimum image in time and depth, respectively, is crucial. A velocity model suitable for depth imaging processes is required for this purpose. However, velocity model building and depth migration in salt bearing basins is a challenge for several reasons. The velocity contrast between the Messinian Evaporites and the overburden is larger than 2 km/s since interval velocities of 4.3 - 4.4 km/s were calculated for the evaporites and 1.7 - 2.1 km/s for the overburden, respectively (Netzeband et al., 2006a). In the depth migration, this strong velocity contrast leads to smearing effects as it cannot be included correctly by the required smooth description of the velocity field.

The data example covers 2000 CMP locations and was acquired in a 2D manner. The total line length is ~ 25 km. The record length is 8 s with 4 ms sample rate and shot and receiver spacing is 25 m. 576 channels were employed yielding a maximum offset of 7325 m. For the purpose of data reduction two adjacent channels were stacked during conventional preprocessing. The acquisition parameters are also listed below.

Line length ~25 km	Acquisition type end-on
Recording time 8 s	Time sample rate 4 ms
Maximum CMP fold 280	Source-receiver offset range [0 m – 7325 m]
Number of CMPs 2000	CMP spacing 12.5 m

The associated CRS stacked section is displayed in 6.2. The section shows salt roller structures and the Messinian Evaporites, which are folded, causing strong lateral velocity contrasts at their top. Thus, these data are suitable to investigate both developed methods.

6.2 Reflection imaging

Time imaging

In this section, I demonstrate the applicability of the developed reflection-imaging workflow for the time imaging which includes an automatic update of time migration velocities. In the first step, we perform an automatic CMP stack of the data to estimate the stacking velocities. In the second step, we conduct a CSP data mapping, described in chapter 2, with the estimated stacking velocities. In the third step, we apply an automatic CMRE stack to the CSP gathers.

The first item in the workflow is to obtain reliable stacking velocities. We make use of the automatic CMP stack which belongs to tools of the high-density stacking velocity analysis (Mann, 2002). The high-density velocity analysis provides more detailed information about the seismic reflection data compared to the conventional approach with smoothed stacking-velocity models based on selected CMP locations and reflection events. Also, the high-density velocity analysis automatically avoids the undesirable stretch effects that are present in the conventional NMO approach (Perroud and Tygel, 2004). To remove possible fluctuations and outliers, we apply event-consistent smoothing based on a combination of median filtering and averaging (Mann and Duveneck, 2004).

The second step in the workflow is the transformation of CMP gathers into CSP gathers (see chapter 2 for more details). In the third step, we apply an automatic Common-Migrated-Reflection-Element stack to the CSP gathers. The CMRE stack applied to the CSP data approximates the zero-offset reflection traveltime with a second-order Taylor expansion in the vicinity of the image ray (see chapter 3 for more details). The image ray is a ray which starts normal to the measurements surface and travels down to hit the reflector at the image-incident-point (Tygel et al., 2009). There is no directional dependency in the traveltime and, therefore, no reflection point smearing for the dipping reflector in the CSP data. Due to increased reflector resolution of the CSP gathers and the absence of diffractions, application of our workflow to the CSP data provides an improved time-migration velocity model with an enhanced vertical resolution.

Figures 6.3 and 6.4 show a CMP gather (6.3a), the corresponding CSP gather (6.3b) and their velocity spectra (6.4c,d). As mapping velocities I used stacking velocities obtained by the automatic CMP stack during the first step of the time-imaging workflow. The CSP data mapping as well as the time migration were performed with the same velocity model. For the CSP gathers, an automatic CMRE stack was performed. Figure 6.5 shows the prestack time migrated section from CMP data (a) and the CMRE stack (b). The reflections appear more continuous, top of salt and base of salt appear a lot more focused in the CMRE-stacked section in comparison to the conventional prestack time migrated section.

106

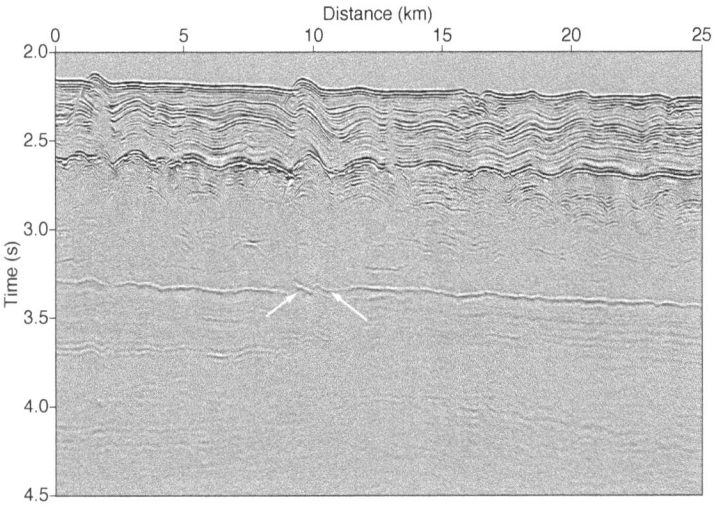

Figure 6.2: Marine data from the Levantine Basin, the automatic CRS stacked section.

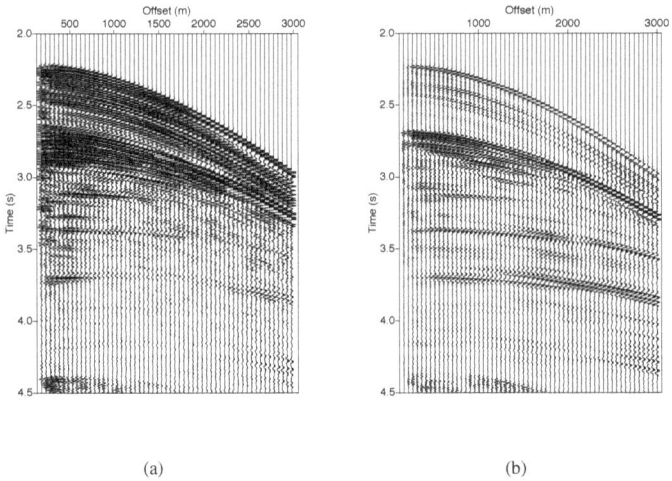

Figure 6.3: Marine data from the Levantine Basin.   CMP gather (a) and corresponding CSP gather (b).

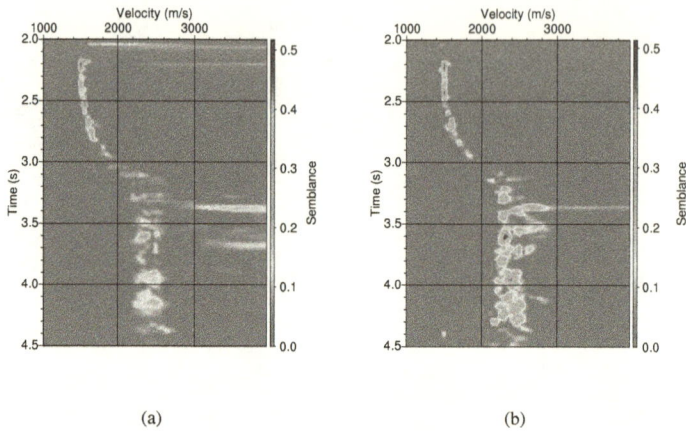

Figure 6.4: Marine data from the Levantine Basin. Velocity spectra of the CMP gather (a) and of the CSP gather (b). The latter has a higher semblance value for prominent events.

Depth imaging

Beside of providing the time-migrated section the CSP data may also be used for several applications. One of the potential applications is velocity model building based on the image ray tomography (see chapter 4). As the time image reveals (Figure 6.5b), the top of salt is visible at 2.6 s, the base of salt is at 3.3 s. The velocity in the sediments is about 2000 m/s and in the salt it is about 4000 m/s. Diffracted events, that occur along the top of salt, indicate a fractured structure of the top of salt and mask the faults. Due to the significant velocity contrast between the sediments and the salt and a complex stratigraphy it is a challenge to reconstruct a detailed velocity distribution for the base of salt. Figure 6.6a illustrates the section of the radii of NIP-wave. Note, this section is weighted with a coherence value of 0.4. The coherence value is of great importance in the reflection tomography because the picking of traveltime information is carried out for events with a determined coherence value. We usually consider that a prominent reflection event has a high coherence value. Especially automatic methods for event picking solely rely on the coherence value and it is often abandoned to the contractor to control the picking.

108

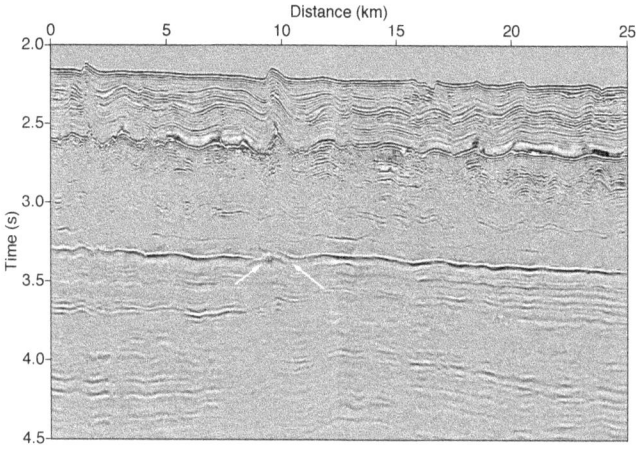

(a)

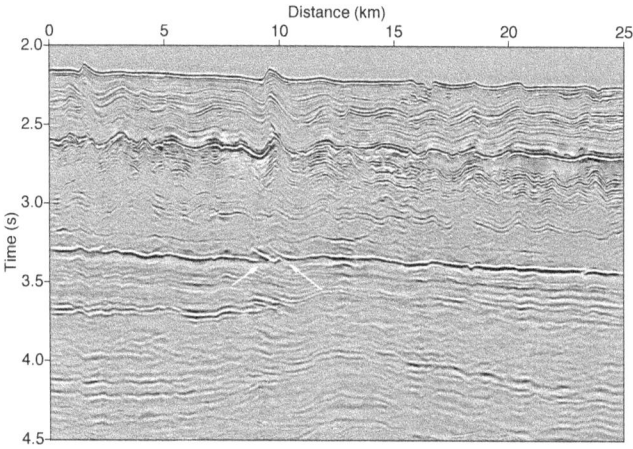

(b)

Figure 6.5: Marine data from the Levantine Basin. Prestack time migrated section (a) and automatic CMRE stack of the CSP gathers (b). Reflections appear more continuous, and top of salt and base of salt are imaged a lot more focused in the CMRE stack section in comparison to the conventional time-migrated section.

Choosing the coherence value quite high reduces the number of events while the usage of lower values runs the risk to pick non-physical events. We see in Figure 6.6a that we were not able to pick a significant amount of reliable attributes. Therefore, this section is not suitable to provide a reliable data vector for NIP-wave tomography in an automatic manner.

Figure 6.6b shows the section related to the wavefront radii of the image-rays. The section is also weighted with a coherence value of 0.4. We can observe that much more attributes can be involved in the image ray tomography in comparison to the section to the NIP-wave tomography, especially from the part containing the salt. 935 data points (t^M, M_{IIP}, x) extracted from CSP gathers serve as the input for the presented inversion method. The velocity model consists of 651 B-spline knots: 31 knots in the x-direction with a spacing of 1000m and 21 knots in the z-direction with a spacing of 250m. The initial model is built using a near-surface velocity of 1500 m/s and a vertical velocity gradient of 0.5 s^{-1}. Also, the constraint for a minimum deviation of the velocity was used because of the presence of salt layers in the data. Figure 6.7b shows the result of the image-ray tomography. The base of salt is well resolved. Compared to the result of the NIP wave tomography, the image-ray tomography produces a smoothed velocity model.

The smooth velocity model obtained with the image-ray tomography was used to perform a prestack depth migration. We used a Kirchhoff migration algorithm based on eikonal traveltimes. A number of common-image gathers at selected image locations with a regular separation of 2 km along the section are shown in Figure 6.8a. The displayed offsets range from 0 to 3000 m. Most events in the CIGs are flat indicating that the determined velocity model agrees with the data fairly well. Some residual moveout is visible in the stack of internal salt reflectors around 2 – 4 km depth. The deep reflectors below the base of salt are, however, perfectly flat.

Figure 6.8b shows a stack of prestack depth migration for the offset range between 0 and 3000 m. We observe that the top of salt at $\sim$ 2 km is well resolved. As expected In contrast and really unexpected, the base of salt and a number of steep faults bellow are well resolved. Those parts of the subsurface below $\sim$ 4 km depth correspond to

110

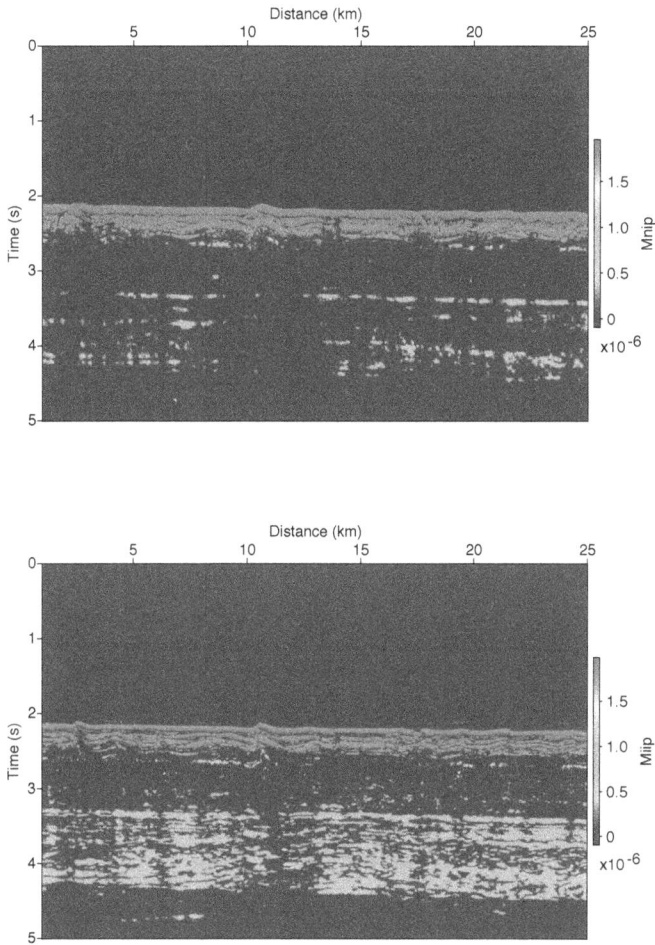

Figure 6.6: Marine field data example. (a) The parameter M_{NIP}, contains information on NIP-wave radii. The parameter M_{IIP} contains information on the wavefront curvature radii of the image rays. The sections are weighted with a coherence value of 0.4. In comparison to the M_{NIP}-section, more picks from the M_{IIP}-section can be involved in the tomographic inversion.

111

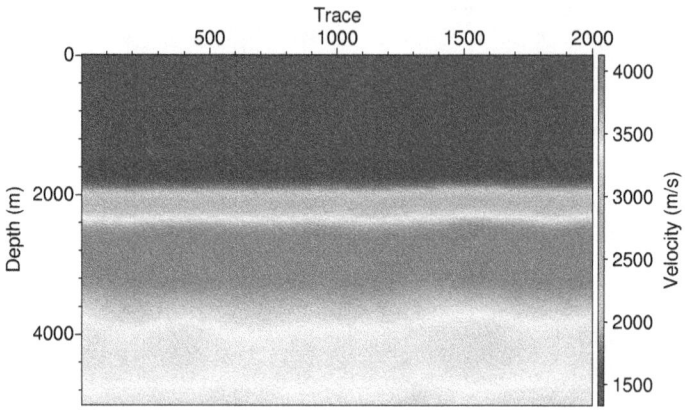

(a)

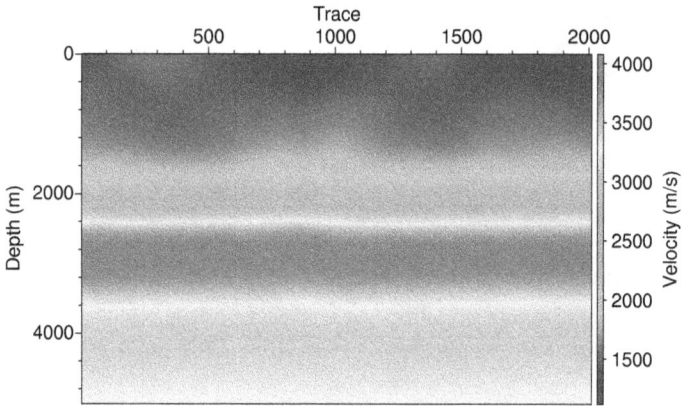

(b)

Figure 6.7: Results of a tomographic inversion with NIP-wave and IIP-waves curvature. The NIP-wave inversion was kindly provided by Stefan Dümmong (University of Hamburg, now at Statoil).

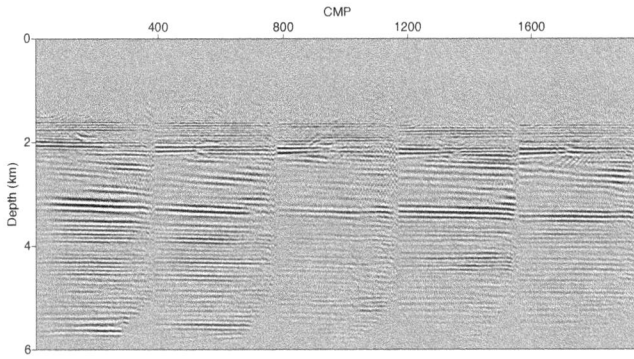

(a)

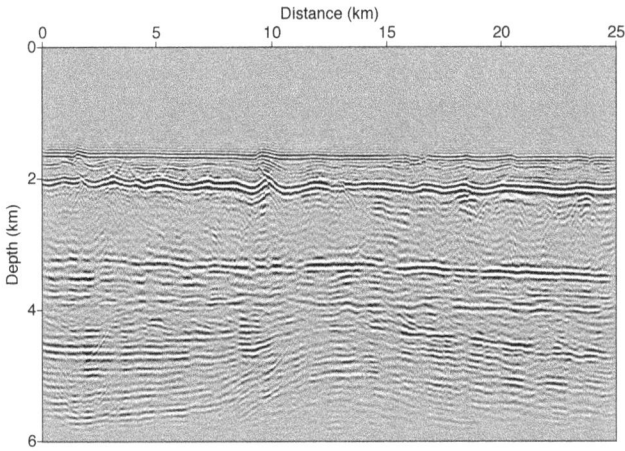

(b)

Figure 6.8: (a) Common Image Gather distributed along the whole profile. The reflected events are almost flat which indicates a good determination of migration velocities. The result of the proposed inversion method can be indeed used as an initial model for migration velocity analysis. (b) Depth migrated section. We can observe a horst and graben along with step faulting can be recognized. Disturbing shallow level strata caused by faulting can be also interpreted. This disturbing is also validated by the presence of many diffractions in the shallow part of the seismic unmigrated section.   113 regions of low coherence in the section containing the NIP-wave attributes (Figure 6.6). Es expected NIP-based velocity model should have a very low resolution for the region. In the salt body reflectors exhibit stronger amplitudes, i.e., strong acoustic impedance contrast, which obviously acts as a barrier for seismic waves traveling downwards. Beneath salt formations we can observe a horst and graben with step faulting. Disturbance shallow salt level strata caused by faulting can also be interpreted. This disturbance is also validated by the presence of many diffractions in the shallow part of the unmigrated seismic section. Salt is clearly seen above the basement rocks which disturb the strata like a bulge. The image of the reflectors beneath the salt body is very crucial for seismic exploration.

We can observe in general that base of salt bottom is very well imaged and also some reflectors below the salt are visible. We can also claim that small thickness undulations of the evaporite unit represent indeed faults and are not apparent velocity pull-ups/-downs. These regions are indicated with white arrow in Figure 6.2.

6.3 Diffraction imaging

In this section, I demonstrate the applicability of the developed diffraction-imaging workflow. The workflow includes an automatic velocity analysis using separated diffractions. The CRS stacked section of the whole recorded data is shown in Figure 6.2a. Diffracted events that occur along the top of salt confirm its fractured structure. Figure 6.9a shows the stacked section after the diffraction separation. Because of the complexity of the data, we used a low threshold of 0.7 during the filtering. Consequently, more residual reflections are present in the stacked section in comparison to the synthetic examples shown before. However, the reflection events between the seafloor and top of salt are strongly attenuated, leaving well-imaged diffraction events from the top of salt.

The field data example showed higher residual reflected energy compared to the synthetic 2D examples. This may well be a 3D effect of diffractors located transverse to the profile line. Another possible explanation of this observation could be the frequency content of the data. The field data example contains considerably higher frequencies in the signal spectrum than the synthetic data. Diffraction imaging, particularly the separation reflections and diffractions, is frequency dependent. What might appear as a *point diffractor* to a low frequency signal may well be a "reflecting horizon" for a high frequency signal. The corresponding Fresnel volume defines the limits of resolution here. The frequency content of the data thus highly influences the filter performance. The same threshold in the filter process will not provide the same performance with respect to the separation of diffraction and reflection events if the frequency content of the data differs.

I applied time-migration velocity analysis on the diffraction-only data (Figure 6.10). I then performed poststack Kirchhoff time migration with the estimated velocity model. Figure 6.9b shows the time-migrated image of the diffraction-only data. The diffractions are focused and indicate the rugged structure of the salt.

Combining both time images improves the overall image of the subsurface.

On the final migrated sections we can observe that the Messinian Evaporites are overprinted by plate tectonics. The presence of many diffractions produced by the top of salt indicates the gravity gliding. Also, the presence of the internal layers in the salt and bulge-like disturbance of the strata by the basement rocks point out the salt tectonics. To sum up, I conclude all three mechanisms, i.e., salt tectonics, gravity gliding, and plate tectonics should be considered for correct formation analysis of the Messinian Evaporites.

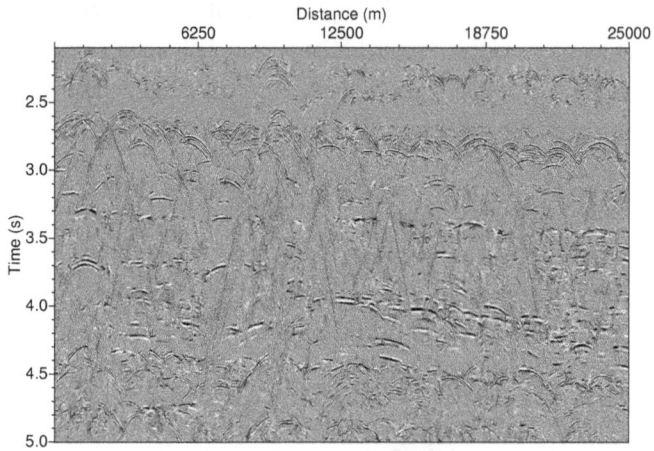

(a)

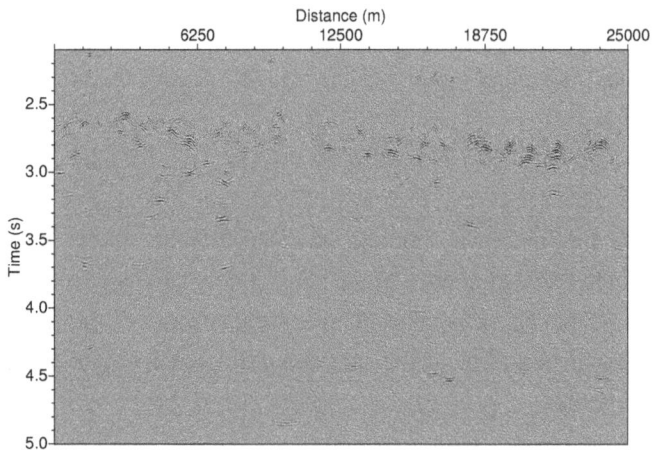

(b)

Figure 6.9: Marine data from the Levantine Basin. (a) CRSD stacked section of diffraction-only data. Reflections are still present in (b), because of the lower threshold we have chosen for the threshold function. (b) Time-migrated section obtained with the velocities estimated from diffractions. The diffractions are focused and indicate the rough topography of the top of salt.

116

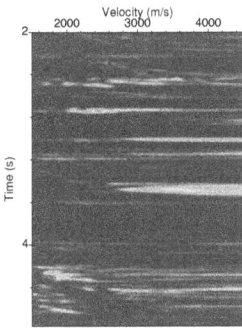
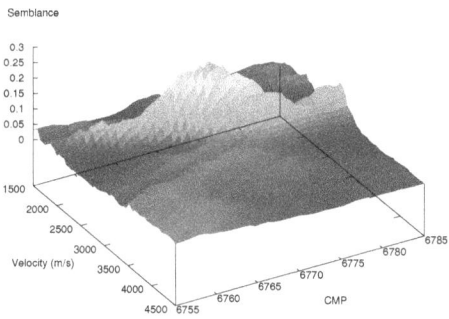

Figure 6.10: Marine data from the Levantine Basin. (a) Velocity spectrum after poststack time migration velocity analysis of a CMP located at 3400 m directly above the diffractor. (b) Coherence of this CMP where the time was manually picked and corresponds to the maximum of the coherence value at 2.87 s in (a). The semblance is considerably smaller compared to the synthetic examples. A distinct maximum can be still identified.

7 Conclusions

In the scope of this thesis, I have studied transformations of seismic data from one domain into another one according to the type of the recorded wavefield. I have focused on two major types: reflection and diffraction events of the recorder wavefield. Since the central step in the seismic imaging still represents reflection imaging, the main part of the thesis is dedicated to the transformation of the seismic data into reflection-only data. For this purpose, I have developed a method that maps the prestack seismic data into 'partly' prestack time-migrated data by exploitation of Kirchhoff's diffraction approach. The developed method generates common scatterpoint, CSP, gathers which maintain the moveout. A CSP gather collects all scattered energy from a 3-D data volume (m,h,t) within the migration aperture and redistributes this energy into a 2-D data volume (h,t) along a hyperbolic path. If a scatter point is exactly at the output location of a CSP gather, the scattered energy will be stacked constructively. Energy from scatter points offset from the output location is canceled through destructive interference. The CSP data mapping uses the parametrization of the DSR equation with the apex time. The summed amplitude is directly mapped into the CO apex of the migration operator. The time domain formulation of the data mapping allows CSP gathers to be formed at arbitrary locations, i.e., a regularized gather is obtained. The stacking in this process provides an improved S/N ratio in the CSP gathers compared to the CMP data.

Since the CSP data represent 'pure' reflections, these data are most suitable for conventional processing steps like stacking or velocity analysis. I have applied a multiparameter stacking method to CSP data. The method is based on the automatic common-migrated-reflector-element, CMRE, stack. The CMRE stack is based on the Taylor expansion of time-migrated traveltimes in the vicinity of the central image ray and its application focuses CSP traveltimes at zero-offset. Since the CSP data mapping requires initial migration velocities, which might be incorrect, the velocity updates are a very important step in the whole procedure. During the CSP data mapping we recalculate the time-migration velocity for every sample, i.e., every CMP coordinate, and every offset, in order to find the one which best fits the traveltime. Therefore, the CSP data mapping contains an integrated velocity update which is the first update. Then, after CSP gathers are generated, I apply the CMRE stack. The CMRE stack belongs to the high-density velocity analysis tools including automatic searches for three stacking parameters. The first search is the automatic CSP stacking. During this search we determine a best-fit stacking parameter which represents the updated time-migration velocity. This search is the second update. During the CMRE stack we also determine best-fit stacking parameters in the midpoint direction. This allows us to account for lateral velocity changes. The application of a subsequent simultaneous parameters optimization provides the final update of time-migration velocities. However, if the initial migration velocity for computing the CSP data is substantially wrong, the CMRE stacking procedure will only improve focusing of the time-migrated reflections similar to the RMO analysis. The CMRE stacking cannot correct a potential damage of a CSP gather and put reflections at their correct positions. Therefore the limitations of the method are related to complex velocity models with strong lateral velocity variations.

CSP gathers can be used for complementary applications, e.g., for multiple suppression or velocity model building. I have also investigated its ability to build smooth velocity models for depth imaging using the wavefront curvature of the image ray. I have proposed to extract these curvatures from the time-migrated reflections since they are related to image rays in a similar way as normal incidence rays relate to primary reflections. The time-migrated primary reflections can be obtained by tracing image rays vertically down from the surface to the considered reflector in depth. The distinct advantage of using kinematic information extracted from the time-migrated reflections lies in the fact that the time-migrated data are free from diffractions and triplications. The other advantage is that the time-migrated data exhibit an improved signal-to-noise ratio. This allows us to determine the kinematic information from the data even in the presence of a low signal-to-noise ratio and conflicting dip situations in the unmigrated data, where identifying reflection events becomes more difficult. I have proposed to build velocity models in a tomographic fashion. The data vector for the inversion contains wavefront curvatures of the image ray. The model vector is calculated by dynamic ray-tracing along central image rays. The inversion problem is solved iteratively by computing the least-squares solution to the locally linearized problem. The Fréchet derivatives required for the tomographic matrix are calculated with ray perturbation theory.

The other major part of the seismic data is represented by diffractions. Diffractions are still seen as the *step child* of the seismic processing. However, they carry detailed information on the subsurface in regions of high importance for the reservoir characterization and exploitation. They provide a naturally and physically justified way to the high-resolution image beyond the classical Rayleigh limit. Moreover, the diffracted wavefield is determined solely by properties of the medium in a small neighborhood of the scatterer. Therefore, diffractions can be used to extract detailed velocity information in the nearest vicinity of the scatterer providing an illumination usually superior to reflections. I have proposed a new method to image seismic diffractions using kinematic wavefield attributes. The radii of the normal wave and NIP wave allow to identify seismic events. For diffractions both radii should coincide, i.e., their ratio is close to one which allows to isolate them from reflections. The imaging consists of stacking of the coherent events with a CRS-based diffraction operator, followed by attenuation of reflected events in the poststack domain with a new type of filter. Finally, the diffracted events are enhanced using partial CRSD stacks. Time-migration velocity analysis of the diffraction-only data followed by Kirchhoff time-migration of the diffractions completes the diffraction imaging.

In the poststack domain, the data are reduced and a good S/N ratio is present. Both facts lead to a reliable and fast time-migration velocity analysis. The developed velocity analysis consists of a velocity scan and the evaluation of the corresponding semblance norm. Applications of the method to synthetic and field data data demonstrate that time-migration velocities estimated from the diffraction-only data lead to a better focused time-migrated image compared to a time migration with RMS velocities. The developed method therefore represents a fast, robust and stable technique for poststack time-migration velocity analysis. I have also investigated in what way diffraction-only data can be used to extract time-migration velocities for reflections. The synthetic data application shows good focusing of the reflections from the layer containing diffractions. Focusing reflections is not sufficient if no diffractions are present in the corresponding layer. However, the field data usually contain a great deal of diffractions randomly distributed along the seismic profile. The velocity information extracted by the presented method from these diffractions can surely be used as an initial velocity model for further velocity analysis using the prestack data-volume.

In general, I do not consider reflection and diffraction imaging as competitors. The transformation of the data into reflection-only or diffraction-only data allows us to adjust seismic-processing tools for particular needs depending on the interpreter's goal and the geological interpretation. The combination of images produced by both techniques will improve the final image.

Suggestions concerning the extensions of the presented methods to anisotropic media are outlined in the following "Outlook" chapter.

8 Outlook

This thesis has introduced techniques for seismic imaging based on diffractions. However, applications of the techniques have been limited to isotropic media since the mathematical framework to describe the diffraction response is valid only for the isotropic case. The major issue for the future will be the extension of the presented techniques to anisotropic media. The last section gives some thoughts on how this could be achieved.

The most frequently used approximation for diffraction traveltimes was derived by Alkhalifah and Tsvankin (1995) and represents an extension of the conventional double-square-root (DSR) equation with a fourth-order term. This term depends on a single additional parameter, η, which steers the non-hyperbolicity of the moveout. The authors assume weak VTI anisotropy. This simplification implies that the vertical time is the fastest one and the traveltime equation is symmetrical with respect to the vertical axes. Unfortunately, this simplification often fails for many cases. Figure 8.1 shows a diffraction traveltime curve for a homogeneous VTI anisotropic medium. The ratio $V_p(0)/V_p(90)$ is equal to two. It is apparent from Figure 8.1 that the fastest time from the diffractor to the surface does not coincide with the vertical time. The fastest ray arrives the measurement surface with an angle of 35 degree. This implies that the apex of the diffraction traveltime curve is laterally shifted with respect to the diffractor location. To properly describe the diffraction response, this shift should be considered. In heterogeneous isotropic media we also observe a shift which need to be considered.

We propose to parametrize the diffraction response with four parameters, the emergence angle of the fastest ray emitted from the diffractor, the ray velocity along this ray and first and second-order derivatives of the ray velocity with respect to the angle. These four independent parameter are solely required to describe the prestack diffraction response in an arbitrary anisotropic heterogeneous medium (Dell et al., 2012b).

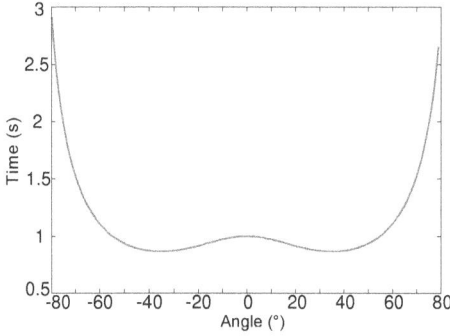

Figure 8.1: The traveltime plot for a point diffractor located in a homogeneous strongly anisotropic VTI medium. The fastest time from the diffractor to the surface does not coincide with the vertical time. The fastest ray arrives at the measurement surface with an angle of about 35 degree. This implies that the apex of the diffraction traveltime curve is shifted with respect to the vertical axis. To properly describe the diffraction response, this shift should be considered.

Also we show that the proposed method provides slowness or phase velocity, respectively, for every point under consideration. The information on the ray and phase velocity can be further used for stacking or an inversion of estimated effective anisotropy parameters into interval anisotropic parameters, in order to apply them, e.g., in lithology discrimination, fracture detection and time-lapse seismic.

Theory

We consider an arbitrary anisotropic inhomogeneous medium with the scheme of observation given in Figure 8.2 where $\bar{x}$ and $\bar{z}$ determine the position of a diffractor in the subsurface, x is the surface projection of the diffractor position, x_0 is a point on the measurement surface belonging to the ray emitted from the diffractor and arriving with the fastest time. x_1, x_2 are coordinates of the collocated source and receiver.

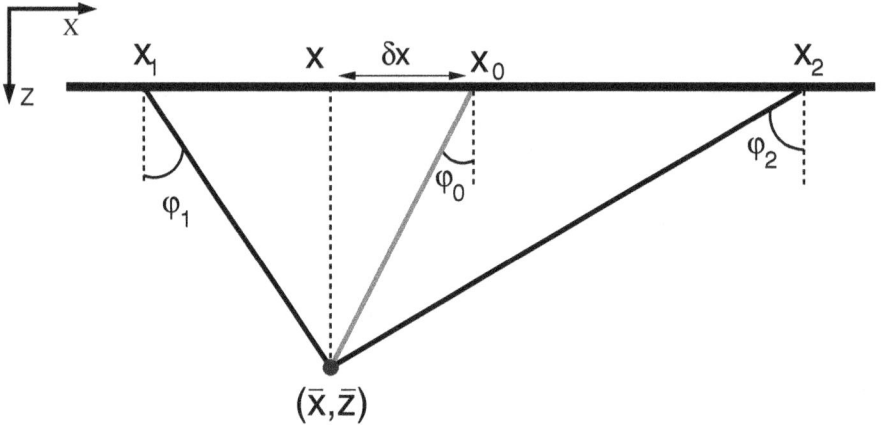

Figure 8.2: Scheme of observation. The travel path is the sum of two legs: $(x_1, 0), (\bar{x}, \bar{z})$ and $(\bar{x}, \bar{z}), (x_2, 0)$. The traveltime can be easily computed using Pythagoras theorem.

The traveltime for a collocated source and receiver, i.e., from $(x_1, z = 0)$ via $(\bar{x}, \bar{z})$ to $(x_2, z = 0)$ is

$$t = \frac{1}{\xi(\varphi_1)} \sqrt{(x - x_1)^2 + \bar{z}^2} + \frac{1}{\xi(\varphi_2)} \sqrt{(x_2 - x)^2 + \bar{z}^2}, \qquad (8.1)$$

where we have introduced angles as shown in Figure 8.2, and ξ represents ray velocity.

Now we expand the ray velocity into a Taylor series in the vicinity of the fastest ray with respect to the angle. For velocities along both branches we obtain

$$\xi(\varphi_1) \approx \xi_0 + \left.\frac{\partial \xi}{\partial \varphi}\right|_{\varphi_0} (\varphi_1 - \varphi_0) + \frac{1}{2} \left.\frac{\partial^2 \xi}{\partial \varphi^2}\right|_{\varphi_0} (\varphi_1 - \varphi_0)^2,$$

$$\xi(\varphi_2) \approx \xi_0 + \left.\frac{\partial \xi}{\partial \varphi}\right|_{\varphi_0} (\varphi_2 - \varphi_0) + \frac{1}{2} \left.\frac{\partial^2 \xi}{\partial \varphi^2}\right|_{\varphi_0} (\varphi_2 - \varphi_0)^2. \qquad (8.2)$$

Denoting

$$A = \frac{1}{\xi_0} \frac{\partial \xi}{\partial \varphi}\bigg|_{\varphi_0} \quad \text{and} \quad B = \frac{1}{2\xi_0} \frac{\partial^2 \xi}{\partial \varphi^2}\bigg|_{\varphi_0},$$

we obtain the following expressions for the ray velocities

$$\xi(\varphi_1) \approx \xi_0\left(1 + A(\varphi_1 - \varphi_0) + B(\varphi_1 - \varphi_0)^2\right),$$

$$\xi(\varphi_2) \approx \xi_0\left(1 + A(\varphi_2 - \varphi_0) + B(\varphi_2 - \varphi_0)^2\right), \qquad (8.3)$$

which we substitute then into equation 8.1

$$t = \frac{\sqrt{(x - x_1)^2 + \bar{z}^2}}{\xi_0\left(1 + A(\varphi_1 - \varphi_0) + B(\varphi_1 - \varphi_0)^2\right)} +$$

$$\frac{\sqrt{(x_2 - x)^2 + \bar{z}^2}}{\xi_0\left(1 + A(\varphi_2 - \varphi_0) + B(\varphi_2 - \varphi_0)^2\right)}. \qquad (8.4)$$

Our approximation depends on the coordinate of the diffractor in the subsurface $(\bar{x}, \bar{z})$, the coordinates of the collocated source and receiver, $(x_1, z = 0)$ $(x_2, z = 0)$, first-order and second-order derivatives of the ray velocity with respect to the ray angle, A and B, angle variations, $\varphi - \varphi_0$, and the ray velocity along the fastest ray, ξ_0. The approximation is not easy to implement in the seismic processing. Therefore, we will establish in the next two sections a time-driven and depth-driven approach to approximate the diffraction response.

Time-driven approximation of the diffraction response

To obtain a time-driven approximation of the diffraction response, we express the x, z coordinate by the fastest time. We calculate the fastest traveltime by simply application of Pythagoras theorem,

$$t_0 = \frac{2}{\xi_0}\sqrt{(x - x_0)^2 + \bar{z}^2} = \frac{2}{\xi_0}\sqrt{\delta x^2 + \bar{z}^2}, \qquad (8.5)$$

125

where $\xi_0 = \xi(\varphi_0)$ is the ray velocity along the fastest ray (red line in Figure 8.2). We can also express the δx in terms of angles. Making use of the sine law yields

$$\delta x = t_0 \xi_0 \sin \varphi_0. \tag{8.6}$$

We expand terms $(x - x_1)$ and $(x_2 - x)$ in Equation 8.1 as followed

$$t = \frac{1}{\xi(\varphi_1)} \sqrt{(x_0 - x_1)^2 - 2(x_0 - x_1)\delta x + \delta x^2 + \bar{z}^2 +} \\ \frac{1}{\xi(\varphi_2)} \sqrt{(x_2 - x_0)^2 + 2(x_2 - x_0)\delta x + \delta x^2 + \bar{z}^2}, \tag{8.7}$$

and make use of equation 8.5. We obtain

$$t = \frac{1}{\xi(\varphi_1)} \sqrt{(x_0 - x_1)^2 - 2(x_0 - x_1)\delta x + \frac{t_0^2 \xi(\varphi_0)^2}{4} +} \\ \frac{1}{\xi(\varphi_2)} \sqrt{(x_2 - x_0)^2 + 2(x_2 - x_0)\delta x + \frac{t_0^2 \xi(\varphi_0)^2}{4}}. \tag{8.8}$$

Extracting $\xi(\varphi_0)$ leads to

$$t = \frac{\xi(\varphi_0)}{\xi(\varphi_1)} \sqrt{\frac{t_0^2}{4} + \frac{(x_0 - x_1)^2}{\xi(\varphi_0)^2} - \frac{2(x_0 - x_1)\delta x}{\xi(\varphi_0)^2}} + \\ \frac{\xi(\varphi_0)}{\xi(\varphi_2)} \sqrt{\frac{t_0^2}{4} + \frac{(x_2 - x_0)^2}{\xi(\varphi_0)^2} + \frac{2(x_2 - x_0)\delta x}{\xi(\varphi_0)^2}}. \tag{8.9}$$

We can invert equations given by Formula 8.3 as

$$\frac{1}{\xi(\varphi_1)} \approx \frac{1}{\xi_0} \left(1 - A(\varphi_1 - \varphi_0) + (A^2 - B)(\varphi_1 - \varphi_0)^2 \right), \\ \frac{1}{\xi(\varphi_2)} \approx \frac{1}{\xi_0} \left(1 - A(\varphi_2 - \varphi_0) + (A^2 - B)(\varphi_2 - \varphi_0)^2 \right). \tag{8.10}$$

and substitute them into Equation 8.9. This provides

$$t \approx \sqrt{\frac{t_0^2}{4} + \frac{(x_0 - x_1)^2}{\xi(\varphi_0)^2} - \frac{2(x_0 - x_1)\delta x}{\xi(\varphi_0)^2}} \times$$
$$\left(1 - A(\varphi_1 - \varphi_0) + (A^2 - B)(\varphi_1 - \varphi_0)^2\right) +$$
$$\sqrt{\frac{t_0^2}{4} + \frac{(x_2 - x_0)^2}{\xi(\varphi_0)^2} + \frac{2(x_2 - x_0)\delta x}{\xi(\varphi_0)^2}} \times$$
$$\left(1 - A(\varphi_2 - \varphi_0) + (A^2 - B)(\varphi_2 - \varphi_0)^2\right). \tag{8.11}$$

We can relate angle and displacement using trigonometrical formulas, i.e.,

$$\tan \varphi_1 = \frac{x_1 - x}{\bar{z}},$$
$$\tan \varphi_2 = \frac{x_2 - x}{\bar{z}},$$
$$\tan \varphi_0 = \frac{x - x_0}{\bar{z}}. \tag{8.12}$$

Furthermore, we will use an approximation of the tangent up to the second order, i.e.,

$$\tan y \approx y + \frac{1}{3}y^3.$$

Figure 8.3 shows an error plot between tangents function and its arguments.

For small angles, the tangent is approximately equal to its argument, i.e., $\tan y \approx y$. Therefore we can establish the following equalities

$$\varphi_1 - \varphi_0 \approx \frac{x_1 - x_0}{\bar{z}},$$
$$\varphi_2 - \varphi_0 \approx \frac{x_2 - x_0}{\bar{z}}. \tag{8.13}$$

The tangents approximation is a limitation in the formula. The approximation is best for small angles/offsets. To avoid this limitation, Dell et al. (2012b) have recently

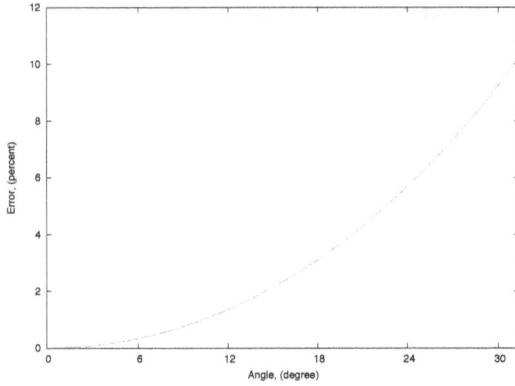

Figure 8.3: Error plot for tangents approximation of the angle, i.e., $\tan y \approx y$. For small angles, tangents function is approximately equal to its argument.

proposed to replace the tangent approximation with a sine-based approximation for the ray-angle difference.

Also we can express the shift δx and the depth $\bar{z}$ by the angle of emergence, i.e.,

$$\delta x = t_0 \xi_0 \sin \varphi_0 ,$$
$$z = t_0 \xi_0 \cos \varphi_0 . \tag{8.14}$$

Substituting equalities given by Equation 8.13 and 8.14, and denoting that $x_0 - x_1 = x_m - h$ and $x_2 - x_0 = x_m + h$ yields an expression which represents the traveltime approximation of the diffraction response for an arbitrary inhomogeneous anisotropic medium. The equation reads

$$t = \sqrt{\frac{t_0^2}{4} + \frac{(x_m - h)^2}{\xi(\varphi_0)^2} - \frac{2(x_m - h)t_0 \sin \varphi_0}{\xi(\varphi_0)}} \left(1 - C(x_m - h) + D(x_m - h)^2\right) +$$
$$\sqrt{\frac{t_0^2}{4} + \frac{(x_m + h)^2}{\xi(\varphi_0)^2} + \frac{2(x_m + h)t_0 \sin \varphi_0}{\xi(\varphi_0)}} \left(1 - C(x_m + h) + D(x_m + h)^2\right) , \tag{8.15}$$

where h is the half offset, and x_m is the midpoint coordinate of the fastest ray, and C

128

and D are determined as

$$C = \frac{A}{t_0 \xi_0 \cos \varphi_0} = \frac{1}{t_0 \xi_0^2 \cos \varphi_0} \frac{\partial \xi}{\partial \varphi},$$

$$D = \frac{A^2 - B}{t_0^2 \xi_0^2 \cos^2 \varphi_0} = \frac{\left(\frac{\partial \xi}{\partial \varphi}\right)^2 - \frac{\xi_0}{2} \frac{\partial^2 \xi}{\partial \varphi^2}}{t_0^2 \xi_0^4 \cos^2 \varphi_0}. \tag{8.16}$$

Inspecting this formula we see that it depends on four parameters. The parameter D contains information on the wavefront curvature. To show this, we reformulate parameter D as

$$D = \frac{D'}{t_0^2 \xi_0^4 \cos^2 \varphi_0}. \tag{8.17}$$

The curvature of the wavefront of the fastest ray at the point x_0 is given as

$$k = \frac{\xi_0^2 + 2\left(\frac{\partial \xi}{\partial \varphi}\right)^2 - \xi_0 \frac{\partial^2 \xi}{\partial \varphi^2}}{\left[\sqrt{\xi_0^2 + \left(\frac{\partial \xi}{\partial \varphi}\right)^2}\right]^3} = \frac{\xi_0^2 + 2D'}{\left[\sqrt{\xi_0^2 + \left(\frac{\partial \xi}{\partial \varphi}\right)^2}\right]^3}.$$

From here

$$D' = \frac{k}{2}\left[\sqrt{\xi_0^2 + \left(\frac{\partial \xi}{\partial \varphi}\right)^2}\right]^3 - \frac{\xi_0^2}{2}. \tag{8.18}$$

Substituting Equation 8.18 in Equation 8.17 leads to

$$D = \frac{\frac{k}{2}\left[\sqrt{\xi_0^2 + \left(\frac{\partial \xi}{\partial \varphi}\right)^2}\right]^3 - \frac{\xi_0^2}{2}}{t_0^2 \xi_0^4 \cos^2 \varphi_0}. \tag{8.19}$$

Equation 8.19 shows that the anisotropic stacking parameter D is directly related to the curvature of the wavefront measured at the acquisition surface.

The new traveltime approximation given by Equation 8.15 is valid for both hetero-geneous and anisotropic media. Now we investigate several special cases:

Isotropic homogeneous media:

For an isotropic homogeneous medium, the first and second-order derivatives are equal zero and the ray velocity coincides with the phase velocity. Also the fastest ray coincides with the vertical ray so that the angle of emergence, φ_0, is zero degree. In this case, our traveltime approximation given by Equation 8.15 simplifies to

$$t = \sqrt{\frac{t_0^2}{4} + \frac{(x_m - h)^2}{v^2}} + \sqrt{\frac{t_0^2}{4} + \frac{(x_m + h)^2}{v^2}}, \tag{8.20}$$

where v is now the phase velocity. The traveltime depends on one stacking parameter, the phase velocity v. The given expression is identical to the conventional DSR operator.

Isotropic heterogeneous media:

In such media, the first and second-order derivatives are equal to zero and the ray velocity coincides with the phase velocity. However, the fastest ray may not coincide with the vertical ray because of inhomogeneities. The traveltime approximation, Equation 8.15, simplifies to

$$t = \sqrt{\frac{t_0^2}{4} + \frac{(x_m - h)^2}{v^2} - \frac{2(x_m - h)t_0 \sin \varphi_0}{v}} +$$
$$\sqrt{\frac{t_0^2}{4} + \frac{(x_m + h)^2}{v^2} + \frac{2(x_m + h)t_0 \sin \varphi_0}{v}}. \tag{8.21}$$

The traveltime depends on two stacking parameters, the phase velocity, v, and the angle of emitting, φ_0. The inhomogeneity is hidden in the third term of the roots which can be interpreted as a lateral shift of the diffraction apex with respect to the

130

diffractor location.

Weakly anisotropic VTI media:

In the case of weak VTI anisotropy, the fastest ray coincides with the vertical ray, i.e., $\varphi_0 = 0$. Also the first derivative of the ray velocity with respect to the ray angle is equal to zero (Tsvankin, 2001). The traveltime approximation, Equation 8.15, simplifies

$$t = \sqrt{\frac{t_0^2}{4} + \frac{(x_m - h)^2}{\xi(\varphi_0)^2}} \left(1 + D(x_m - h)^2\right) + \sqrt{\frac{t_0^2}{4} + \frac{(x_m + h)^2}{\xi(\varphi_0)^2}} \left(1 + D(x_m + h)^2\right).$$

$$(8.22)$$

The traveltime depends on two stacking parameters, ray velocity ξ, and the parameter D. It can be shown that for weak VTI $D \approx -\delta$. Note, δ is here the Thomsen parameter. The present approximation of the diffraction response depends also on two parameters: vertical P-velocity V_{p_0} and a combined parameter η (Alkhalifah, 2000).

Estimation of the slowness/phase velocity

The fastest ray arrives at the measurement surface at point x_0 with a ray angle equal to φ_0 and the ray velocity ξ. Since the traveltime of the fastest ray should be minimum, the horizontal projection of the slowness vector p_x equals zero, i.e.,

$$\left.\frac{\partial t}{\partial x}\right|_{x_0} = 0.$$

This also implies that the phase angle θ equals zero degree. The 2-D Eikonal equation

$$(\xi, \mathbf{p}) = \xi_x p_x + \xi_z p_z = 1,$$

131

simplifies to

$$\xi_z p_z = 1 \, ,$$

leading to

$$p_z = \frac{1}{\xi_z} \, .$$

Thus if for every point (x_0, t_0) the vertical component of the ray velocity ξ_z is determined, the vertical component of the phase velocity is also determined.

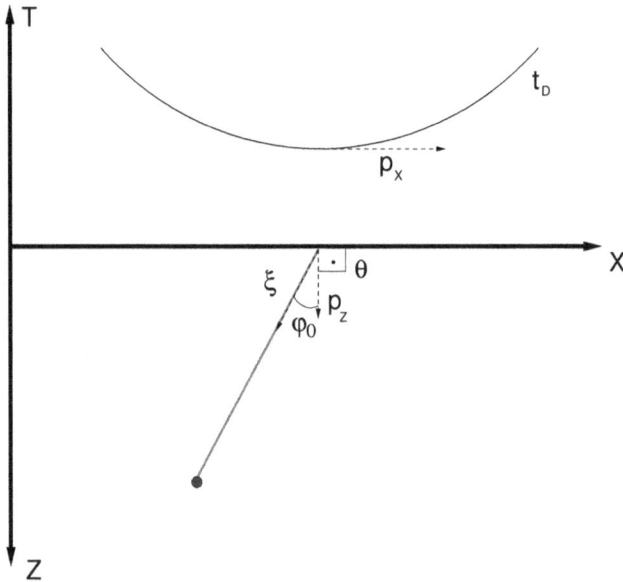

Figure 8.4: A cartoon illustrates how to estimate the slowness/phase velocity.

Depth-driven approximation of the diffraction response

For some applications, e.g., localization of passive seismic events, the time-driven approximation of the diffraction response is not sufficient. In this section we will derive a depth-driven approximation of diffraction response. For this purpose, we directly substitute Equations 8.13 in Equation 8.4. The traveltime approximation reads

$$t = \frac{\bar{z}^2}{\xi_0} \frac{\sqrt{\bar{z}^2 + (x_m - h)^2 - 2\delta x(x_m - h) + \delta x^2}}{\bar{z}^2 + A\bar{z}(x_m - h) + B(x_m - h)^2} +$$
$$\frac{\bar{z}^2}{\xi_0} \frac{\sqrt{\bar{z}^2 + (x_m + h)^2 + 2\delta x(x_m + h) + \delta x^2}}{\bar{z}^2 + A\bar{z}(x_m + h) + B(x_m + h)^2}. \tag{8.23}$$

We see the depth-driven approximation depends on the same parameters as the time-driven approximation.

For the VTI media we obtain after some simplifications the following equation

$$t = \frac{1}{\xi_0} \sqrt{\bar{z}^2 + x^2} \frac{1}{1 + A\frac{x}{\bar{z}} + B\frac{x^2}{\bar{z}^2}}. \tag{8.24}$$

We see that the anisotropy is hidden in the fraction term. The term will become less than one with an increasing distance to the diffractor. This leads that the traveltime correction will converge to the isotropic case for $x \gg z$. This corresponds to horizontal propagation, i.e only horizontal velocity need to be considered. This conclusion coincides with general observations (Tsvankin, 2001).

For weak VTI equation 8.24 simplifies to

$$t = \frac{1}{\xi_0} \sqrt{\bar{z}^2 + x^2} \frac{1}{1 + \delta\frac{x^2}{\bar{z}^2}}, \tag{8.25}$$

where δ is the Thomsen parameter (Thomsen, 1986).

List of Figures

Bibliography

Aki, K. and Richards, P. (1980). *Quantitative Seismology*. Freeman, San Francisco.

Alkhalifah, T. (2000). The offset-midpoint traveltime pyramid in transversely isotropic media. *Geophysics*, 65(4):1316–1325.

Alkhalifah, T. and Tsvankin, I. (1995). Velocity analysis for transversely isotropic media. *Geophysics*, 60:1550–1566.

Bancroft, J., Geiger, H., and Margrave, G. F. (1998). The equivalent offset method of prestack time migration. *Geophysics*, 63:2042–2053.

Baykulov, M. and Gajewski, D. (2009). Prestack seismic data enhancement with partial common-reflection-surface (CRS) stack. *Geophysics*, 74:no. 3, V49–V58.

Berkovitch, A., Belfer, I., Hassin, Y., and Landa, E. (2009). Diffraction imaging by multifocusing. *Geophysics*, 74:no. 6, WCA75–WCA81.

Berryhill, A. W. (1977). Diffraction response for nonzero separation of source and receiver. *Geophysics*, 42:1158–1176.

Billette, F. and Lambaré, G. (1998). Velocity macro-model estimation by stereotomography. *Geophysical Journal International*, 135(2):671–680.

Bortfeld, R. (1989). Geometrical ray theory: Rays and traveltimes in seismic systems (second-order approximation of the traveltimes). *Geophysics*, 54:342–340.

Cameron, M. K., Fomel, S. B., and Sehian, J. A. (2007). Seismic velocity estimation from time migration. *Inverse Problems*, 23:1329–1369.

Cartwright, J. and Jackson, M. (2008). Initiation of gravitational collapse of an evaporitec basin margin: The Messinian saline giant, Levant Basin, eastern Mediterranean. *Geological Society of America Bulletin*, 120:399–413.

Chernyak, V. S. and Gritsenko, S. A. (1979). Interpretation of the effective common-depth-point parameters for a three-dimensional system of homogeneous layers with curvilinear boundaries. *Geologiya i Geofizika*, 20:112–120.

Dell, S. Gajewski, D. (2011). Common-reflection-surface-based workflow for diffraction imaging. *Geophysics*, 76(5):S187;doi:10.1190/geo2010–0229.1.

Dell, S. and Gajewski, D. (2011). Image Ray Tomography. In *73th Conference and Technical Exhibition, EAGE*, page P148. Extended Abstracts.

Dell, S. and Gajewski, D. (2012). Diffraction imaging in three dimensions and migration velocity analysis. *Geophysical Prospecting*, revised.

Dell, S., Gajewski, D., and Vanelle, C. (2012a). Prestack time migration by common-migrated-reflector-element stacking. *Geophysics*, 77(3):doi:11.1190/geo2011–0200.1.

Dell, S., Pronevich, A., Kashtan, B., and Gajewski, D. (2012b). Diffraction traveltime approximation for anisotropic media. *Geophysics*, submitted.

Dell, S., Tygel, M., and Gajewski, D. (2010). Image-ray tomography. *Geophysical Prospecting*, submitted.

Dix, C. H. (1955). Seismic velocities from surface measurements. *Geophysics*, 20(1):68–86.

Dümmong, S. (2010). *Seismic data processing with an expanded Common Reflection Surface workflow*. PhD thesis, University of Hamburg.

Duveneck, E. (2004). *Tomographic determination of seismic velocity models with kinematic wavefield attributes*. Logos Verlag Berlin.

Farra, V. and Madariaga, R. (1987). Seismic wavefront modeling in heterogeneous media by ray perturbation theory. *Geophys. Res.*, 92:2697–2712.

Ferber, R. G. (1994). Migration to multiple offset and velocity analysis. *Geophysical Prospecting*, 42:99–112.

Fomel, S., Landa, E., and Taner, M. T. (2006). Posstack velocity analysis by separation and imaging of seismic diffractions. *Geophysics*, 72:no. 6, U89–U94.

Geiger, H. D. (2001). *Relative-amplitude-preserving prestack time migration by the equivalent offset method*. PhD thesis, University of Calgary.

Gelchinsky, B. (1982). Scattering of waves by a quasi-thin body of arbitrary shape. *Geophysical Journal of the Royal Astronomical Society*, 71:915–928.

Gradmann, S., Hübscher, C., Ben-Avraham, Z., Gajewski, D., and Netzeband, G. (2005). Salt tectonics off northern Israel. *Marine and Petroleum Geology*, 22:597–611.

Hagedoorn, J. (1954). A process of seismic reflection interpretation. *Geophys. Prosp.*, 2:85–127.

Harlan, W. S., Claerbout, J. F., and Rocca, F. (1984). Signal/noise separation and velocity estimation. *Geophysics*, 49:1869–1880.

Hertweck, T., Schleicher, J., and Mann, J. (2007). Data stacking beyond CMP. *The Leading Edge*, 26(7):818–827;DOI:10.1190/1.2756859.

Hubral, P. (1975). Locating a diffractor below plane layers of constant interval velocity and varying dip. *Geophysical Prospecting*, 23:313–322.

Hubral, P. (1977). Time migration – Some ray theoretical aspects. *Geophysical Prospecting*, 25:738–745.

Hubral, P. (1983). Computing true amplitude reflections in a laterally inhomogeneous earth. *Geophysics*, 48:1051–1062.

Hubral, P. and Krey, T. (1980). *Interval velocities from seismic reflection traveltime measurements*. Soc. Expl. Geophys.

Hubral, P., Schleicher, J., and Tygel, M. (1992). Three-dimensional paraxial ray properties -I. Basic relations. *Journal of Seismic Exploration*, 1:265–279.

Hübscher, C. and Netzeband, G. (2007). Evolution of a young salt giant: The example of the Messinian evaporites in the Levantine Basin. *In: Wallner, M.,*

Lux, K.-H., Minkley, W., Hardy, Jr., H.R. (Eds.) The Mechanical behaviour of Salt - Understanding of THMC Processes in Salt, pages 175–184.

Iversen, E. and Tygel, M. (2008). Image-ray tracing for joint 3D seismic velocity estimation and time-to-depth conversion. *Geophysics*, 73, n.3:S99–S114.

Jäger, R. (2004). *Minimum aperture Kirchhoff migration with CRS stack attributes.* PhD thesis, University of Karlsruhe.

Jäger, R., Mann, J., Höcht, G., and Hubral, P. (2001). Common-reflection-surface stack: Image and attributes. *Geophysics*, 66:97–109.

Kanasewich, E. R. and Phadke, S. M. (1988). Imaging discontinuities on seismic sections. *Geophysics*, 53:334–345.

Keller, J. B. (1962). Geometrical theory of diffraction. *Journal of the Optical Society of America*, 52:116–130.

Khaidukov, V., Landa, E., and Moser, T. J. (2004). Diffraction imaging by focusing-defocusing: An outlook on seismic superresolution. *Geophysics*, 69:1478–1490.

Klokov, A., Baina, R., and Landa, E. (2010). Separation and Imaging of Seismic Diffractions in Dip Angle Domain. In *72th Conference and Technical Exhibition, EAGE*, page G040. Extended Abstracts.

Klüver, T. and Mann, J. (2005). Event-consistent smoothing and automated picking in CRS-based seismic imaging. In *75th Annual International Conference, SEG*, pages 1894–1897. Expanded Abstracts.

Kozlov, E., Barasky, N., Koroloev, E., Antonenko, A., and Koshchuk, E. (2004). Imaging scattering objects masked by specular reflections. In *74th Annual International Conference, SEG*, pages 1131–1135. Expanded Abstracts.

Kravtsov, Y. A. and Ning, Yan., Z. (2010). *Theory of diffraction. Heuristic approaches.* Alpha Science.

Krey, T. (1952). The Significance of Diffraction in the Investigation of Faults. *Geophysics*, 17:843–858.

Kunz, B. F. J. (1960). Diffraction problems in fault interpretation. *Geophysical Prospecting*, 8:381–388.

Landa, E. and Keydar, S. (1997). Seismic monitoring of diffraction images for detection of local heterogeneities. *Geophysics*, 63:1093–1100.

Landa, E., Shtivelman, V., and Gelchinsky, B. (1987). A method for detection of diffracted waves on common-offset sections. *Geophysical Prospecting*, 35:359–373.

Mann, J. (2002). *Extensions and Applications of the Common-Reflection-Surface Stack Method*. Logos Verlag Berlin.

Mann, J. and Duveneck, E. (2004). Event-consistent smoothing in generalized high-density velocity analysis. *SEG expanded abstracts*, 23:2176.

Martins, J. L., Schleicher, J., Tygel, M., and Santos, L. (1997). 2.5-D True-amplitude Migration and Demigration. *Journal of Seismic Exploration*, 6:159–180.

Mayne, W. H. (1962). Common reflection point horizontal data stacking techniques. *Geophysics*, 27:927–938.

Moser, T. J. and Howard, C. B. (2008). Diffraction imaging in depth. *Geophysical Prospecting*, 56:627–641.

Müller, N.-A. (2007). *Determination of interval velocities by inversion of kinematic 3D wavefield attributes*. PhD thesis, TH Karlsruhe.

Neev, D. (1975). The Pelusium Line - a major transcontinental shear. *Tectonophysics*, 38:T1–T8.

Neev, D. (1977). Tectonic evolution of the Middle East and the Levantine basin (easternmost Mediterranean)T. *Geology*, 3:683–686.

Neev, D., Almagor, G., Arad, A., Ginzburg, A., and Hall, J. (1976). The geology of the Southeastern Mediterranean Sea. *Geological Survey of Israel Bulletin*, 68:1–51.

Netzeband, G., Gohl, K., Hübscher, C., Ben-Avraham, Z., Dehgahni, A., Gajewski, D., and Liersch, P. (2006a). The Levantine Basin - crustal structure and originMessinian evaporites in the Levantine Basin. *Tectonophysics*, 418:178–188.

Netzeband, G. L., Hübscher, C. P., and Gajewski, D. (2006b). The structural evolution of the Messinian evaporites in the Levantine Basin. *Marine Geology*, 230:249–273.

Perroud, H. and Tygel, M. (2004). Nonstretch NMO. *Geophysics*, 69:599–607.

Popov, M. M. (2002). *Ray Theory and Gaussian Beam Method for Geophysicists*. EDUFBA, Salvador Bahia.

Reshef, M. and Landa, E. (2009). Post-stack velocity analysis in the dip-angle domain using diffractions. *Geophysical Prospecting*, 57:811–821.

Robein, E. (2010). *Seismic Imaging. A Review of the Techniques, their Principles, Merits and Limitations*. EAGE Publications BV.

Sava, P. C., Biondi, B., and Etgen, J. (2005). Wave-equation migration velocity by focusing diffractions and reflections. *Geophysics*, 70:no. 3, U19–U27.

Schleicher, J., Tygel, M., and Hubral, P. (1993). Parabolic and hyperbolic paraxial two-point traveltimes in 3D media. *Geophysical Prospecting*, 41:495–513.

Schleicher, J., Tygel, M., and Hubral, P. (2007). *Seismic True-Amplitude Imaging*. SEG monograph.

Spinner, M. (2007). *CRS-based minimum-aperture Kirchhoff migration in the time domain*. PhD thesis, University of Karlsruhe.

Tarantola, A. (1987). *Inverse problem theory:Methods for data fitting and model parameter estimation*. Elsevier, Amsterdam.

Thomsen, L. (1986). Weal elastic anisotropy. *Geophysics*, 51:1954–1966.

Tsvankin, I. (2001). *Seismic signatures and analysis of reflection data in anisotropic media*. Pergamon.

Tygel, M., Mueller, T., Hubral, P., and Schleicher, J. (1997). Eigenwave based multiparameter traveltime expansions. In *67th Annual International Conference, SEG*, pages 1770–1773. Expanded Abstracts.

Tygel, M., Ursin, B., Iversen, E., and de Hoop, M. V. (2009). An interpretation of CRS Attributes of time-migrated reflections. *WIT Reports*, Annual Report No. 13:260–268.

Tygel, M., Ursin, B., Iversen, E., and de Hoop, M. V. (2010). Depth conversion of zero-offset and time-migrated reflections. *WIT Reports*, Annual Report No. 13:210–224.

Ursin, B. (1982). Quadratic wavefront and traveltime approximations in inhomogeneous layered media with curved interfaces. *Geophysics*, 47:1012–1021.

Vanelle, C. (2002). *Traveltime-Based True-Amplitude Migration*. PhD thesis.

Vanelle, C. and Gajewski, D. (2002). Second-order interpolation of traveltimes. *Geophysical Prospecting*, 50:73–83.

Červený, V. (2001). *Seismic ray theory*. Cambridge University Press, Cambridge.

Waterman, P. C. (1976). Matrix theory of elastic wave scattering. *Journal of the Acoustical Society of America*, 60:1086–1097.

www.ingramcontent.com/pod-product-compliance
Lightning Source LLC
Chambersburg PA
CBHW021103210326
41598CB00016B/1305